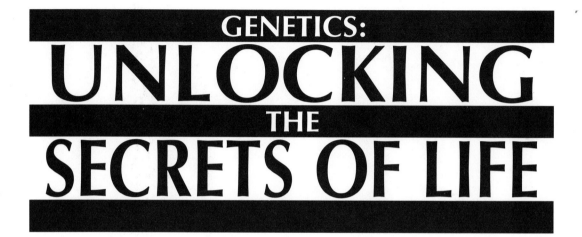

GENETICS: UNLOCKING THE SECRETS OF LIFE

GENETICS:
UNLOCKING
THE
SECRETS OF LIFE

Nathan Aaseng

The Oliver Press, Inc.
Minneapolis

The Oliver Press, Inc.
Charlotte Square
5707 West 36th Street
Minneapolis, MN 55416-2510

The publisher wishes to thank **Dr. Harley Geller,**
Cedars-Sinai Medical Center, Los Angeles, California,
for his careful review of this manuscript.

Library of Congress Cataloging-in-Publication Data
Aaseng, Nathan.
Genetics : unlocking the secrets of life / Nathan Aaseng.
p. cm. — (Innovators)
Includes bibliographical references and index.
   Summary: Examines the lives and scientific
breakthroughs of seven scientists who contributed to the
field of genetics—Charles Darwin, Gregor Mendel, Thomas
Hunt Morgan, Oswald Avery, James Watson, Francis Crick,
and Har Gobind Khorana.
ISBN 1-881508-27-7
1. Geneticists—Biography—Juvenile literature.
[1. Geneticists.] I. Title. II. Series.
QH437.5.A27 1996
575.1'09—dc20   95-38804
                                                  CIP
                                                  AC

ISBN: 1-881508-27-7
Innovators II
Printed in the United States of America

02 01 00 99 98 97 96   7 6 5 4 3 2 1

CONTENTS

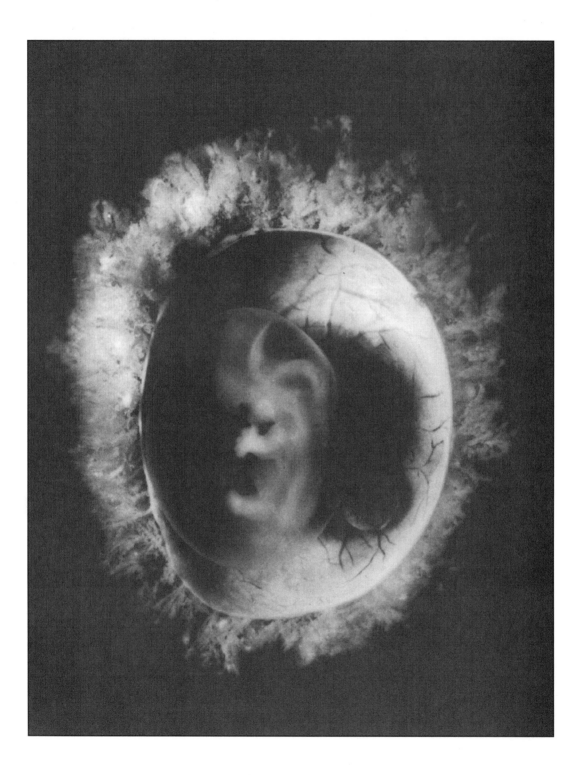

The Seeds of Knowledge

Genetics, the study of how characteristics are passed from one generation to the next, is a relatively new science. The word *genetics*, which comes from a Greek word meaning "to give birth to," was first used at the beginning of the twentieth century. Since that time, genetics has come to dominate the field of biology. Secrets uncovered by geneticists hold both the promise and the threat of bringing about radical changes in our world—from the elimination of disease to the test-tube creation of entirely new forms of life.

Although the science of genetics is only a century old, the basic questions of genetics have been on people's minds ever since the first humans looked at the world around them. Two facts about the world must have been obvious to even humanity's most distant ancestors. The first was that life displays an incredible amount of variety. Plants can range in size from a fraction of an inch to a towering 100

The study of genetics has answered many basic questions, such as why this developing human baby, a 42-day-old fetus within the amniotic sac, might have blue eyes like her father. Geneticists are still searching for answers to more complex questions, such as the recent discovery of pattern-forming genes that direct the growth and organization of the human embryo.

species: a kind or variety of similar living things that are able to breed together

feet above the ground. They can be good to eat or poisonous, and they can display flowers of almost every shade in the rainbow. Animals can range from tiny worms to massive elephants, from aquatic jellyfish to great blue whales.

The second fact was that despite this impressive variety, offspring tend to look like their parents. Seeds from pine trees do not produce plants that look like moss. Cats do not give birth to babies that resemble a rhinoceros. Even among members of the same species, parents tend to share with their offspring such characteristics as eye and skin color, height, and facial features.

Genetics asks how this happens. What causes creatures to inherit these two opposite conditions of variety and sameness? How is likeness passed on to a new generation? And if only likeness is passed on, how is variety created?

The earliest explanations of inheritance had more to do with religion than with science. For centuries, people believed that the process of inheritance could not be understood by humans.

The Greeks who lived 2,500 years ago, however, were not content to call heredity a supernatural mystery. Many early Greek scholars believed that the world was an orderly place where things happened for a reason. The Greeks tried to understand why children would look like their parents. They knew that mating was involved in producing offspring. Was there some element of sexual union that transferred traits from parents to their offspring?

The Greek physician Hippocrates believed so. He taught that the "seed material" for creating a

Ancient people knew that a pine cone could only grow into a pine tree, but the reasons why have only become understood in the last 100 or so years.

baby came from all parts of both parents' bodies. During fertilization, the mother's seed material mixed with the father's. Because this material, possibly in particle form, contained a tiny bit of every part of the body, it could grow into a human. The child would be a combination of all the physical characteristics of both parents. The mixture of two parents' traits explained why children resembled both parents but did not look exactly like either one.

Aristotle, a Greek philosopher who wrote about plants and animals in the fourth century B.C., followed a similar line of thinking. He believed that parents passed on characteristics to offspring through a mixture of their elements that was collected through the blood. (Aristotle's influence on heredity continues to this day. People still commonly refer to "blood lines" when speaking of their ancestry and to "blood relations" to indicate those to whom they are most closely related genetically.) But Aristotle proposed a difference between the male and the female seed elements. The female, he said, provided the unformed raw material out of which a person was made—what he called "matter." The male's seed element, the "form," had the power to shape that raw material into the finished product.

For the next 2,000 years, no one produced a more persuasive explanation about how traits passed from generation to generation than Aristotle's. One of the barriers to understanding heredity was a widespread acceptance of spontaneous generation. While aware that sexual intercourse was involved in creating life, many people believed that new life could

Hippocrates (460-357 B.C.) taught that medicine should be based on observation and science, not on superstition. This theory earned him the title of "the father of medicine."

Greek philosopher Aristotle (384-322 B.C.) believed that characteristics were passed on through the blood.

also be created without parents. According to the concept of spontaneous generation, new creatures could suddenly appear from nonliving substances. For example, maggots could spring from rotting meat. Because spontaneous generation required no parents, there was no point in looking for a parental mechanism for passing along traits.

In Europe during the Middle Ages (about 400-1400) and the Renaissance (1400-1600), heredity was primarily a religious issue. The Christian world believed that God had created the earth and everything in it. God's plan had arranged for the separate creation of all forms of life, and each life was an individual act of creation. That God had chosen to have offspring resemble parents was just another part of the divine mystery. In the seventeenth century, a Dutchman named Jan Swammerdam (1637-1680) went so far as to argue that all the seeds of every living creature had been created when God formed the world.

By the seventeenth century, individuals probing into the workings of life discovered that all female mammals had ovaries. These ovaries produced eggs that developed into living creatures. Late in the same century, Antony van Leeuwenhoek verified that these eggs required sperm from the male in order to produce offspring.

The discovery that both parents contributed visible elements to the birth process supported the ancient Greek explanation of heredity as a mixture of material from both the mother and the father. This led to more speculation as to how these parts

Antony van Leeuwenhoek (1632-1723) was one of the first to use a microscope to study microorganisms.

produced new life. Leeuwenhoek thought that the sperm contained an entire, fully formed, new creature and that the female egg provided a suitable environment in which this creature could grow. The English physician William Harvey, who described the developing stages of an embryo, believed that all living creatures come from an egg.

At this point, scientists in France began to dominate the subject of genetics. In the eighteenth century, French mathematician Pierre de Maupertuis (1698-1759) became curious about members of a particular family who had six fingers or six toes on

In addition to his theories of heredity, English physician William Harvey (1578-1657) also described how the heart pumped, circulating blood through the body.

each hand or foot, instead of five. He traced this unusual trait through four generations of the family and concluded that particles from both parents were responsible for the makeup of the offspring.

During the eighteenth century, the discovery of fossils raised many questions about the creation and development of life on earth. These fossils appeared to be remains of extinct forms of life, but religious scholars, who were often the people most involved in the study of nature, were puzzled about why God would allow so many creations to be destroyed. They also struggled to explain why the older rocks contained only fossils of simpler animals while the more complex types of animals appeared only in the more recent rocks. Most of them explained these fossil finds with the theory of catastrophism. This theory held that periodically some natural event, such as a flood or an ice age, annihilates some species, which are later replaced by other species. However, French scientists (who were less likely than their colleagues in other European countries to be

A fossil is formed when the remnant of a plant or animal becomes imbedded and preserved in the earth's crust. Fossils, like this calamite or horsetail, forced eighteenth-century scientists to rethink accepted ideas about the origins of life.

intimidated by the teachings of the church) looked for alternative explanations to the one proposed by those who believed in catastrophism. Commenting on the great variety in the world, Georges de Buffon proposed that living things did not necessarily stay as God had originally created them but underwent changes over the years.

In the early nineteenth century, Buffon's student, Jean-Baptiste Lamarck, carried this line of thought further. Lamarck was a brilliant scientist who combined the study of plants and animals into a single field, which he called biology. He explained the fossil record by proposing that nature had been on the move ever since creation. He thought of the natural world as a staircase in which all living things strived to become more complex. As creatures advanced up the scale in complexity, the lower stairs were emptied. These empty slots were then filled by the spontaneous generation of more simple creatures.

How did creatures become more complex? Lamarck explained this by theorizing that creatures improved themselves by reacting to their environment. Furthermore, parents could pass on these improvements to their offspring. For example, long legs were a useful adaptation for a bird that waded in the water. Certain birds developed these longer legs in response to this environmental need and then passed on this trait to their offspring.

Lamarck was far from the first to propose that offspring could inherit characteristics, and even habits, that parents had acquired during their lives. Even the Greeks had believed that a person's heroic

Georges Louis Leclerc comte de Buffon (1707-1788) advanced the idea that the earth was as old as 75,000 years, and had resulted from a collision between the sun and a comet. This was a daring view at a time when most people believed that the earth and all of life were created whole and at once 6,000 years ago.

While classifying species, Jean-Baptiste Lamarck (1744-1829) was the first naturalist to use the terms "vertebrate" for creatures with spinal columns and "invertebrate" for those without.

qualities were passed along from one generation to the next. But Lamarck was the first to suggest that these acquired changes gradually shaped members of a species, or a specific type of animal, into a new species. "After a long succession of generations," wrote Lamarck, "individuals, originally belonging to one species, became at length transformed."

Lamarck's ideas met with stiff resistance, especially from religious leaders. Based on their interpretation of the Bible, these authorities declared that the earth had existed for only a few thousand years—certainly not long enough to produce even a small amount of gradual variation in creatures.

Advances in one field of science often cause explosive changes in another. The fuse for a biological and genetic revolution was lit when geologists, who study the origin, history, and structure of the earth, began to find evidence that the earth was far older than anyone had earlier imagined.

This book will trace our understanding of how variety and likeness are passed on in nature: from the discovery that species are continually changing in response to the world around them; to the discovery that living things pass on traits to the next generation in the form of chemical instructions; and, finally, to the ingenious ways in which researchers use this information to create new and altered forms of life.

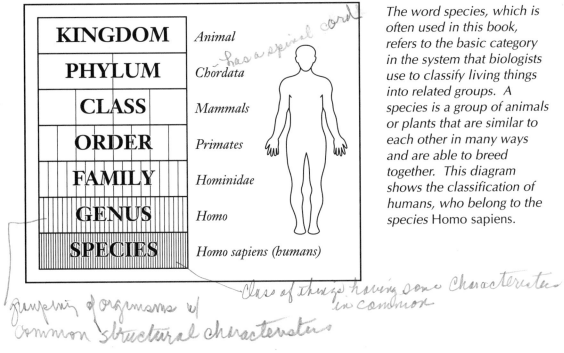

KINGDOM	Animal
PHYLUM	Chordata
CLASS	Mammals
ORDER	Primates
FAMILY	Hominidae
GENUS	Homo
SPECIES	Homo sapiens (humans)

has a spinal cord

The word species, which is often used in this book, refers to the basic category in the system that biologists use to classify living things into related groups. A species is a group of animals or plants that are similar to each other in many ways and are able to breed together. This diagram shows the classification of humans, who belong to the species Homo sapiens.

grouping of organisms w/ common structural characters

Class of things having some characteristics in common

CHAPTER ONE

Charles Darwin and Natural Selection

Perhaps no scientist in history has created such an uproar or has so altered the way in which people look at the world as Charles Darwin. Yet few who knew him would have believed that he was marked either for greatness or controversy. Those were qualities that belonged to his grandfather, Erasmus Darwin.

Holding strong opinions and wading fearlessly into controversy, Erasmus Darwin had built a solid reputation and a personal fortune with his skill as a physician. He challenged church authorities and English public opinion by writing a book in which he boldly supported Jean-Baptiste Lamarck's views that the environment could influence how species evolve.

Charles seemed to have learned nothing from his grandfather. (He later claimed that he did not gain a single idea from reading his grandfather's book.) The young Charles idled away most of his school years and could not summon the will to work

A quiet, deliberate man who avoided arguments, Charles Darwin (1809-1882) created much controversy over his theory of how species evolve.

Erasmus Darwin (1731-1802) was asked by King George III to come to London and be his personal physician, but Darwin declined.

Across the Atlantic Ocean, in far humbler circumstances, another man destined for greatness was also born on February 12, 1809: Abraham Lincoln.

at either of his first two chosen careers—medicine and the ministry. Charles's father despaired that his son would never amount to anything.

Charles did not impress anyone with his quick mind; in fact, he was slow and cautious in both his thoughts and actions. He was so shy and humble that he hid his most important work for 20 years and published it only when he learned that someone else was about to publish a similar paper.

Born in Shrewsbury, England, on February 12, 1809, Charles was the fifth of six children. His mother, Susannah, was the daughter of a famous and wealthy English potter, Josiah Wedgwood. She suffered from bad health and died when Charles was eight. Charles spent most of his growing years under the stern and towering influence of his father, Robert Darwin. Robert, a successful physician like his father, Erasmus, had both a build and manner that were intimidating. He stood six feet, two inches and weighed 350 pounds. When he came home from work in the evenings, he would lecture his children for hours on various subjects.

The young Charles loved roaming through the woods and countryside, collecting plants and animals, and considered classwork an interference with these more important pursuits. He frequently referred to school as "boring" and "a waste of time."

Charles had begun his formal education at a Unitarian day school in 1817. The following year, he enrolled in the Shrewsbury Grammar School. Although the Darwin home was only a mile away, Charles and one of his brothers lived at the school as boarders. Charles put forth little effort in class, angering the schoolmaster and provoking more lectures from his father.

Robert Darwin, however, did notice his son's keen interest in studying living things, and he hoped that this interest might suit the boy to the family profession of medicine. In 1825, he arranged for 16-year-old Charles to attend Edinburgh University in Scotland, which was the finest medical school in Europe at the time.

Charles, however, displayed no more enthusiasm for the university than he had for any other school. Instead, he spent his time socializing, hunting, and exploring the nearby woods and seashores. Edinburgh was then a center for marine biology, and Charles collected and dissected sea creatures. When he witnessed an operation performed without anesthesia to kill the pain, the soft-hearted Darwin realized he did not have the stomach to be a physician. After observing Charles's poor performance in his studies for two years, his father arrived at the same conclusion.

Because the Darwins were wealthy, Charles did not have to earn a living. Nevertheless, his inability to find a useful way to occupy his time embarrassed his father. Searching for some respectable career for his son, Robert settled on the ministry. Unable to oppose his father in any matter, Charles went along with this plan and headed off to Christ's College at Cambridge in 1828.

As before, Darwin studied little and entertained himself as much as possible, spending his

While a student at Cambridge, Darwin wrote to a friend: "I suppose you are two fathoms deep in mathematics and if you are, then God help you, for so am I, only with this difference, I stick fast in the mud at the bottom and there I shall remain."

naturalist: a person who studies and describes organisms and natural objects, especially their origins and interrelationships

father's money freely. Although he later admitted he should have been ashamed of himself, he remembered this as the most joyful time of his life. While not enthusiastic about his religious studies, Darwin did find professors who shared his interest in outdoor subjects. He often went on field trips with botany professor John Stevens Henslow and geologist Adam Sedgwick and impressed both of them with his meticulous collection of specimens and his recording of data. Henslow encouraged him to become a naturalist.

Despite his lack of interest in classroom studies, Darwin completed his requirements at the college in 1831. Although Darwin was content (or, perhaps, resigned) to become a clergyman, shortly before he was to commit to this career, John Stevens Henslow recommended him for an exciting adventure. The HMS *Beagle*, commanded by Captain Robert Fitzroy, was about to set sail on an exploratory trip to South America. The captain was looking for a naturalist to observe and collect specimens of plant and animal life.

While the opportunity thrilled Charles, his father was not supportive. Suspecting that this adventure would reflect badly on Charles's character and make it harder for him to settle down, Robert refused to give his permission. Fortunately, his uncle, Josiah Wedgwood (son of Josiah, the potter), heard about the opportunity and pleaded on Charles's behalf. Robert reluctantly allowed Charles to go.

The *Beagle* set sail on December 27, 1831. Darwin immediately got seasick and never quite

adjusted to the sea. But, two months later, when the ship reached Bahia, Brazil, its first stop on the continent of South America, he began his duties with enthusiasm. Free from the control of his father and the drudgery of the classroom, Darwin eagerly set about doing what he most enjoyed in life—hunting and collecting animals. He climbed steep mountains and explored lush jungles. According to his records, he shot and collected 80 different species of birds in a single morning. Darwin crammed the *Beagle* so full of samples, each neatly documented and sorted, that the crew nicknamed him the "Flycatcher." They joked that Darwin seemed intent on capturing every living thing in South America and sending them all back to England.

The HMS Beagle was a small ship, 90 feet in length, with 74 people on board. "The absolute want of room is an evil that nothing can surmount," Darwin wrote to his professor, John Stevens Henslow.

Charles Lyell (1797-1875) was knighted in 1848 for his important theory explained in The Principles of Geology *(1830). He was working on the book's twelfth edition when he died.*

While carrying out his duties as a naturalist, Darwin was also on the lookout for evidence that supported or refuted a new geological theory proposed by Charles Lyell. In his book, Lyell argued that the earth was far older than most people realized. The world, said Lyell, could not be the same as it had been at its original creation because the earth's surface was constantly changing through such natural processes as erosion and earthquakes.

As the earth changed, the habitat—or surroundings—to which a species was best suited often disappeared. Lyell believed it was this phenomenon, and not a series of devastating catastrophes, that could explain why a species became extinct.

With his keen eye for detail, Darwin noted many curious facts that supported Lyell's beliefs. After a tremendous earthquake rocked the west coast of South America, Darwin saw places where the land had been permanently elevated in the harbor of Concepción, Chile. This proved to him that the beds of mussel shells he had discovered 12,000 feet above sea level in the Andes Mountains had once been submerged in the sea, but had been elevated in more recent geological times by events such as the earthquake. Lyell was right—the earth was indeed changing and altering habitats. Darwin saw that the rock layer in which he had discovered the fossil remains of an extinct sloth also contained many creatures that still existed. That indicated that the sloth was not wiped out by a giant flood or similar catastrophe. For if such a calamity had occurred, the other creatures also would have become extinct.

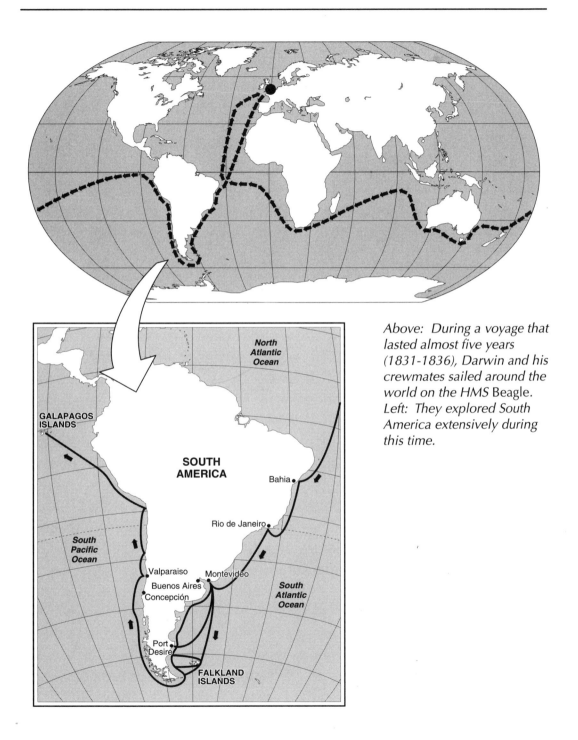

North
Atlantic
Ocean

GALAPAGOS
ISLANDS

SOUTH
AMERICA

Bahia

Rio de Janeiro

South
Pacific
Ocean

Valparaiso
Buenos Aires
Concepción
Montevideo

South
Atlantic
Ocean

Port
Desire

FALKLAND
ISLANDS

Above: During a voyage that lasted almost five years (1831-1836), Darwin and his crewmates sailed around the world on the HMS Beagle. *Left: They explored South America extensively during this time.*

extinct: no longer existing or living

fossil: the remains of a plant or animal from a past geological age that became buried and preserved in the earth's crust

habitat: the environment, or surroundings, in which an organism normally lives

Darwin collected interesting clues that he hoped could shed some light on the subject of how animals adapt, or adjust, to their habitats. On his travels, he observed that the same types of plants tended to grow in the same kinds of places. To him, this indicated that plants were either designed to fit their environment or were influenced by their environment. But then he came across an exception to this rule. The plants he found on the west side of the Andes Mountains were far different from the plants on the eastern slopes of the mountains, even when the growing conditions were similar. What could explain this?

Darwin discovered that many fossils of extinct species were similar to living animals. If species became extinct because of gradually changing habitats, why did some creatures survive while very similar animals did not? Darwin found the most startling patterns of life when the *Beagle* landed on the Galapagos Islands, a bleak collection of small, stony chunks of land 650 miles west of Ecuador. These desolate islands harbored a surprising variety of finches, lizards, and tortoises, most of which were new to Darwin and different from those on the mainland. Furthermore, geological evidence showed that these islands were formed long after animals were thriving on the mainland. In order to produce separate species on the old mainland and on the new islands, God would have had to perform separate acts of creation. After further observation, Darwin also noted that each of the islands was home to its own unique species of animals.

Again, Darwin was puzzled by his observations. The islands were only a few miles apart, shared the same climate, and were made up of the same black rock. Given such identical conditions, why did the animal species differ? If God had designed animals specifically to fit their environment, why put different creatures on identical islands?

While Darwin was working in the Galapagos Islands, the years of feverish, detailed collecting and observing began to take their toll on him, and the naturalist longed to go home. Not until October 2, 1836, nearly two years behind the *Beagle's* original schedule, did Darwin finally return to England. He was so changed in appearance that his father barely recognized him.

Although largely desolate piles of lava rock, with vegetation only on the upper slopes, the 13 Galapagos Islands support some unusual wildlife, such as elephant tortoises. Here Darwin measures a tortoise's walking speed.

In this room at Down House, Darwin did much of his scientific research. The revolving table was crowded with bottles and fossil specimens. He would sit in a low chair to work with the microscope (under the glass jar on the window ledge).

THE BREAKTHROUGH

After arriving home, Darwin immediately set about the task of identifying and cataloging his collection from South America. While doing so, he pondered the ideas of extinction and species variation and adaptation. What produced this constant change and tremendous variety in the world?

Darwin's observations on his voyage convinced him that Lyell was right—the earth was very old and natural processes were constantly at work reshaping it. He agreed with Lyell that changing environments, which caused otherwise well-adapted species to become extinct, was a better explanation of the fossil record than the idea of a series of worldwide catastrophes wiping out many entire species.

But Lyell's theory left an important question unanswered. If species were constantly becoming extinct, why did so many species remain on earth? Were new species being created? If so, how? Could species actually change into other species, as Lamarck had proposed?

At this time, most church and scientific authorities insisted that species could not change. They held firm to the belief that only God could create new life and that God had formed each species separately. What God had made, they declared, was perfect and unchanging.

Darwin was not the sort of person who liked to challenge authority. In fact, during the early months of his voyage on the *Beagle*, the crew made fun of him for rigidly arguing in favor of the religious point

Charles Darwin as he looked when he was completing his theory of evolution

of view in so many discussions. But Darwin could not ignore what his observations showed. He saw that many existing animals were well adapted to their habitat. If the earth was constantly changing, then only two possible explanations existed for why modern animals were well suited to their environments. Either God kept creating new animals or animals were changing along with the earth.

When Darwin looked for evidence that animals could change, he needed to look no further than the work of animal and plant breeders. By choosing parents with desirable traits who could pass on those traits to their offspring, breeders had produced vastly different animal varieties, for example, fast, long-legged greyhounds and stocky, powerful bulldogs. Selective breeding created new strains that eventually became permanent.

Was there, then, evidence that such adaptations occurred in nature? Darwin noticed a pattern in the way species resembled each other. Species on closely neighboring islands were far more similar to each other than to the species found on the mainland. Species on one side of the Andes Mountains were more similar to each other than to species on the other side. Species in Australia, the continent most isolated from other continents, were far different from animals found anywhere else.

Darwin saw that isolation was a key factor in determining how closely species resembled each other. But if creatures were separately created and unchanging, natural barriers should not have any effect. Darwin reasoned that in a population of inbreeding animals, any changes or adaptations would tend to be shared with all the animals of a population. If two groups of one species became isolated from each other, however, they would not share their adaptations and would tend to evolve in their own separate ways. Eventually, they would become so different that they would be separate species.

While this sounded reasonable to Darwin, he was stumped by the question of what shaping force directed that change. In animal breeding, a breeder selected the desired traits. What or who in nature played the role of selecting what new traits or adaptations animals would keep?

In September 1838, Darwin read "Essay on the Principle of Population" by noted British economist Thomas Malthus. Malthus showed that all populations produce far more offspring than could survive. If unchecked, each population would grow until it consumed all of the available resources. But populations do not grow unchecked because they are in fierce competition with other creatures for the same limited resources. The competition becomes so intense that most individuals within a population do not survive to reproduce.

Darwin observed that this struggle for survival occurred not just between rival groups of animals but between animals within a species. An adaptation could give an animal or a plant an advantage, or a greater chance, to survive and to produce more offspring. For example, a bird with a stronger beak than others of its species would be able to crack

The first professor in the field of economics in England, Thomas Malthus (1766-1834) received an appointment as professor of political economy at the East India Company's College at Haileybury in 1805.

tough seeds that the other birds could not eat. This gave it a survival advantage. Such a bird was more likely to live and produce offspring than the bird with a weaker beak. Over time, those that did not have the strong beak might die out, and only those that had the trait of a strong beak would survive.

Adaptations that made an animal more attractive to its mates would also likely result in more offspring that shared such traits. These appealing characteristics would be passed on, and the animals that did not have them would die out. This process could explain why many extinct animals resemble modern animals in so many ways. The extinct creatures simply lacked one or two modifications that the surviving animals had.

Darwin saw this competitive process as the means by which nature caused certain traits to be passed on. He called the process natural selection. Natural selection seemed to support Lamarck's theory that all creatures are moving up the natural staircase on their way to becoming more complex. But Darwin studied his notes and found that the situation was more complicated than Lamarck had stated. Evolution was not a straight line from simple to complex animals, but was more like a tree. Plants and animals developed in different ways, depending on which traits happened to appear and the degree to which groups were isolated from others of their species.

THE RESULT

Darwin reached his conclusions about evolving species and natural selection in 1839. But recognizing the controversy that they would produce, and typically fearful of offending anyone, Darwin moved slowly and cautiously. He did not finish a rough draft of his theory until 1842, and then he spent two more years polishing it.

In 1844, Robert Chambers anonymously published *The Vestige of the Natural History of Creation*, in which he suggested that species were constantly changing. The resulting storm of criticism from scientific and church authorities about the book may have intimidated Darwin, causing him to keep his theory of natural selection private, while continuing to gather more evidence in support of it.

Darwin might never have summoned the courage to publish his work had not a letter and manuscript arrived from Alfred Russel Wallace in June 1858. Wallace, a naturalist working in Southeast Asia, asked Darwin's advice on a theory that he had developed. As Darwin read in horror, Wallace had come up with an explanation of natural selection that was almost identical to the one Darwin had developed more than a decade earlier.

Stunned by the letter and under stress because of family illness, Darwin was ready to surrender any claim to having previously developed the natural selection theory. A generous and courteous gentleman, he could not bear having others suspect him of stealing Wallace's ideas. But after hearing from

Alfred Russel Wallace (1823-1913) had more in common with Darwin than a theory. Like Darwin, he spent his youth searching for a profession and found his calling as a naturalist for a ship. On his voyages, he visited faraway places such as the Amazon basin, the Malay Peninsula, and Australia.

Darwin's friends, with whom Darwin had earlier shared his ideas, the equally polite Wallace gave Darwin credit for coming up with the idea first.

Later in 1858, a group of scientists met in London. At the meeting, friends of Darwin presented papers on natural selection written by both Darwin and Wallace. Overwhelmed by the number of presentations, few listeners understood the impact of what the two naturalists had theorized.

In his usual careful manner, Darwin labored for more than 13 months to compile his ideas into a book. Published on November 24, 1859, his *Origin of Species* created a huge sensation. Many scientists welcomed it and were so astounded by its logic and simplicity that they wondered why no one had thought of these ideas before.

Other readers, however, including religious authorities, expressed shock and outrage at what they read and unleashed a barrage of criticism and ridicule at Darwin. Even many of those who saw the logic of his reasoning fell into despair at having their firmest beliefs about the nature of the world shattered. Some predicted that Darwin's work signaled the end of civilization.

Darwin took no part in the debate. After reporting what he had found, he left the task of arguing the case to others. With his wife, Emma, a cousin whom he had married in 1839, and their children, he retreated into seclusion at their country home near the village of Down. There, with his usual relentless fascination, he continued to explore topics of the natural world.

Whether due to illness or because of inner turmoil over his controversial findings, Darwin lived the rest of his life as a virtual invalid. Some historians think that while on his voyage, he may have contracted trypanosomiasis, an infection caused by a parasite. Suffering from constant aches and pains, he tired easily and napped often. He seldom left the yard of his house or appeared at any social functions, and referred to himself as "a withered leaf."

The complete title of Charles Darwin's book published in 1859 is *On the Origin of Species by Means of Natural Selection, or the Preservation of Favoured Races in the Struggle for Life.* It is usually referred to simply as the *Origin of Species.*

Thomas Huxley (1825-1895) became interested in natural history while sailing to Australia as a ship's surgeon. An influential scientist who served as president of the Royal Society (1881-1885), he ardently lectured and debated in support of Darwin's theory.

To the end of Darwin's life, the workings of the process of inheritance baffled him. Along with Lamarck, he mistakenly believed that people could inherit traits that their parents had acquired during their lives. He guessed that the reproductive cells gradually collected a set of representative particles from all organs and tissues of body. Later insights about this process would have to come from someone else's work.

Darwin's work on natural selection did explain how variety in species gradually became established over many generations. He brought forth the ideas that nature provides a vast pool of random variations and selects those variations that are most useful to pass on to offspring.

When Charles Darwin died on April 19, 1882, scientists of his time acknowledged him as one of the giants of science. His name remains as familiar today as it was then. He was buried in Westminster Abbey among the heroes of England, near Isaac Newton and Charles Lyell.

Although resistance continues to this day, biologists have generally accepted Darwin's work as one of the cornerstones of biology. More importantly, he completely overturned the prevailing notions of the world. More than anyone, Darwin was responsible for separating scientific research from the grip of rigid world views and for freeing scientists to explore the world without fear of offending social and religious authorities.

This photo of Charles Darwin was taken shortly before his death in 1882.

CHAPTER TWO

Gregor Mendel and Dominant and Recessive Traits

While Darwin's argument in favor of natural selection explained most of the available facts about heredity, there remained a few missing pieces that bothered him and his supporters. First, Darwin had no clear explanation as to what caused new traits to appear. How did a bird happen to suddenly develop a large beak that was so efficient in cracking seeds? Darwin did not know enough about the process of inheritance, so he wrongly guessed that animals had some inner mechanism that could produce changes in response to an environment.

Second, like most people, Darwin saw inheritance as a blending of characteristics from the father and the mother. Because blending tended to dilute characteristics, when a bird that somehow had developed a strong beak mated with another bird that had a beak typical for their species, the offspring would have a beak stronger than that of one parent but weaker than that of the other. In the following

Working alone, Austrian monk and botanist Gregor Mendel (1822-1884) received no recognition during his lifetime and never knew that his experiments would provide the foundation for a new branch of science known as genetics.

The ideas of Charles Darwin (1809-1882) and the experiments of Gregor Mendel formed the basis of modern biology.

generation, the trait for a strong beak would be further diluted.

The blending of parental traits meant that any favorable variation appearing in a population was likely to be divided among the group. The favorable variation would diminish and fade back toward the average. In other words, the blending of parental traits actually should lead to less variety among a species, instead of more.

This second missing piece of Darwin's theory of natural selection was found by Gregor Mendel, an Austrian monk. While Darwin's *Origin of Species* was rolling off the presses, Mendel was performing novel experiments that would explain how traits could be passed on independently, with no blending of characteristics. Unfortunately, the few scientists aware of Mendel's experiments ignored them, so Darwin never heard of the man who is considered to be the founder of the field of genetics.

Johann Mendel was born on July 22, 1822, to Anton and Rosina Mendel, who were poor peasants. The Mendels worked a small farm near the village of Heinzendorf, Austria, in a region then known as Moravia, which included parts of the present-day Czech Republic. The Austrian emperor ruled Moravia from Vienna.

Johann shared his father's enthusiasm for outdoor life. Working on a farm gave the boy keen insights into the wonders of plant and animal life. Young Johann was also fortunate to receive an education far superior to that of most poor farm children at the time. Under the guidance of the generous and wealthy Countess Waldburg, the Heinzendorf schools offered a fine program with a special emphasis on science. A teacher named Thomas Makitta taught his students such advanced biological techniques as how to cultivate and create new varieties of fruit, called hybrids, by cross-breeding two different species.

Johann performed well in class, especially in biology. Although Heinzendorf did not have a high school, teachers advised him to continue his education at a nearby town that did have such facilities. Excited by tales of two older boys from his village who had done that, Johann set off in 1833 to attend school at the neighboring village of Troppau.

In 1840, Johann went to the Institute of Olmütz to work toward a two-year degree in philosophy. But in trying to stretch his meager funds so that he could afford his schooling, he practically starved himself and ended up so sick he had to drop out of school for a time. Just as Johann was recovering, a falling tree seriously injured his father, and left him unable to continue farming. Anton Mendel sold his farm and divided the proceeds among his children. Because Johann's inheritance wasn't large enough, his sister Teresia donated her share of the inheritance to help pay for her brother's schooling. So Johann returned to Olmütz and finally earned his degree.

Some historical accounts have portrayed Mendel as a person trained in religious studies who tinkered with plant experiments as a hobby. In reality, most of Mendel's education prepared him for a

biology: the study of life

botany: the study of plants

hybrid: the offspring of two individuals of different species or varieties

career in science. During this time, however, science was not a paying profession because neither industries nor universities employed scientists to perform research. Like Darwin, most scientists came from wealthy families that could support their children while they studied, traveled, and performed experiments. But a young man like Mendel, who came from a poor family, had little opportunity for the leisure of studying science. The only way that Mendel could ever have the time and money to pursue his interest in science was to join the priesthood. Mendel entered the Augustinian monastery at Altbrünn on October 9, 1843, under his chosen name of Gregor, and he was ordained into the priesthood on August 6, 1847.

Although he applied the same discipline to his religious studies as he had to his scientific training, Mendel was not comfortable as a parish priest. Thus, an opportunity to substitute teach at a high school in the town of Znaim in 1849 thrilled him. While he appeared to have taught well at the school, he later failed his official qualifying test to be a teacher. Believing that this was due to a lack of experience and education, in 1851, the Augustine Order sent Mendel to the University of Vienna. There he received instruction from some of the finest scientists of his day.

The extra education failed to salvage his teaching career, for Mendel never did pass the qualifying teacher's examination. The thought of retaking the examination made him so ill (some reports say he had a nervous breakdown) that he never tried again.

Despite his lack of official status, Mendel continued to work part time as a science teacher after returning to Altbrünn. But when the caretaker of the monastery's elaborate, well-tended gardens died, Mendel took over his duties. At last Mendel had found an occupation that he really enjoyed and for which he had obvious skill. Even more importantly for someone with his inventive mind, Mendel was fortunate to have superiors who allowed him to perform his duties in whatever way he saw fit. Freed from stifling supervision and outside pressure, and with the monastery taking care of his basic needs, Mendel now began probing into some questions about which he had long been curious.

THE BREAKTHROUGH

Ever since he had been a young boy on the farm, Mendel had observed how vegetables, grain, fruit, and even animals tended to take after their parents. One of the questions of heredity that had puzzled farmers, animal breeders, and scientists for a long time was why all brothers or sisters didn't look alike. Mixing equal parts of two chemicals or two liquids brought the same result every time. Why then did the mixture of traits from the mother and traits from the father not always yield the same result? Were the hereditary elements not equally mixed?

Franz Unger, one of Mendel's former botany professors, posed another challenge to Mendel. Wrestling with some of the same questions as Darwin about the source of variety in nature, Unger noted that variations in plant characteristics eventually produced new varieties of plants. The process of change, he thought, would continue until new varieties would evolve into new species.

While breeders and professors had often speculated about the nature of heredity, few had actually studied it. Mendel was eager to design some experiments that might finally provide some hard facts. He knew that trying to follow a large number of traits that were able to produce great variety through many generations—such as hair color in humans—would cause confusion. He needed to find subjects that had only a few easily distinguished traits.

After some early experiments with white and gray mice, Mendel settled on a familiar plant, the common garden pea. The pea was ideal for many reasons. It came in varieties with traits that did not mix or blend but rather showed up as either one or the other. For example, there was a tall variety of pea and a dwarf variety. Seeds produced by these plants did not grow into new plants halfway in between the tall and the dwarf but instead tended to be either one or the other.

Peas also grew quickly and in large numbers, and they required a minimum of care. They were self-pollinating and self-fertilizing; that is, they reproduced through the transfer of pollen from the male to the female part of the same flower or of other flowers on the same plant. Therefore, their flowers were not likely to be fertilized by pollen from some unknown plant, which could cause inconclusive results. Finally, peas were useful because, unlike some other plants, the hybrids (seeds produced by cross-pollinating two varieties) remained fertile, or able to reproduce.

Mendel was a meticulous record keeper who for many years had faithfully recorded details of such occurrences as weather and sun spots. Now he brought this same painstaking care to his proposed experiment. First, he obtained 34 varieties of garden peas from a seed dealer. By keeping only those varieties that consistently produced virtually identical offspring when self-fertilized, Mendel narrowed his choices down to 22. From these varieties, he chose pairs that displayed easily recognizable characteristics in two unmistakable forms. He had pairs of dwarf and tall plants; plants that yielded green

The common garden pea proved a useful subject for Mendel's heredity studies because it grew fast and had only a few traits to track.

stamen: the male reproductive part of a flower, which produces pollen containing sperm cells

pistil: the female reproductive part of a flower, which produces ovules containing egg cells

pollination: the transfer of pollen from the stamen to the pistil of a flower

fertilization: the union of a sperm cell and an egg cell. After fertilization takes place, an ovule develops into the seed of a new plant.

seeds and yellow seeds; wrinkled seeds and smooth seeds; and plants whose leaves were terminal, or growing at the end of the stems, and those whose leaves were axial, or growing lined up around the stem.

In 1857, Mendel planted his first rows of peas in the monastery garden. Over the next eight years, he collected the seeds from each plant and recorded their appearance. Then he planted the new seeds and described and pollinated the new plants.

Cross-pollination was the most tedious part of his task. In order to pollinate a pea plant with pollen from another plant, Mendel had to snip open each flower and cut off the male reproductive parts, called stamens, so that pollen would not form.

He then created a new generation of plants by fertilizing the pea with pollen collected from another plant. Finally, he tied cloth bags over each flower to prevent stray pollen from getting in. Mendel also had to be very careful to keep his different varieties and generations of seeds separate.

Working alone, Mendel analyzed thousands of seeds and plants. His data provided him with some interesting facts about heredity in peas. When he allowed his peas to self-fertilize, he found that all of the tall pea plants produced seeds that grew into tall pea plants and that all of the dwarf plants produced seeds that grew into dwarf pea plants.

Since all of the peas had been grown under identical conditions, this proved to Mendel that the height of these pea plants had nothing to do with environmental factors such as soil, water, and temperature. Instead, a hereditary factor must control their height.

When Mendel cross-pollinated tall peas with dwarf peas, the resulting seeds all produced tall pea plants, and the dwarf characteristic disappeared. He realized that heredity was not a simple matter of the equal mixing of elements from two parents. But when Mendel allowed this generation of peas—labeled F1—to self-fertilize, a strange thing happened. The dwarf characteristic reappeared! Mendel repeated this experiment again and again. All cross-pollination between purebred tall and dwarf peas produced an F1 generation of tall peas. Yet whenever this F1 generation of tall peas self-fertilized, it produced an F2 population that included dwarfs.

generation: the interval of time between the birth of the parents and the birth of their offspring

The same thing occurred with regard to seed color, seed shape, and all of the other traits that Mendel was following. All cross-pollination between peas with contrasting traits produced an F1 generation that had only one of those traits. But when this F1 generation self-fertilized, the absent trait reappeared in small numbers in the F2 generation.

Not content with mere observation, Mendel counted and recorded the exact number of each offspring that had been produced. He found, for example, that 253 self-fertilized F1 generation plants produced 5,474 smooth seeds and 1,850 wrinkled seeds. From self-fertilized F1 generation plants, he counted 6,022 yellow seeds and 2,001 green seeds.

Mendel's dogged patience in producing all of these numbers would have counted for little had he not possessed the insight to understand what the statistics meant. He noted that in all cases the ratio of one trait to the other in the F2 generation was very close to 3 to 1.

Suppose, thought Mendel, that peas had two factors that determined heredity for a particular trait. In a purebred tall pea, both factors would favor a tall pea, and in a purebred dwarf pea, both factors would favor a dwarf pea. When the two were combined in pollination, the new plant would get one factor from each parent. The tall factor, however, would override the dwarf. Mendel called the tall factor the dominant trait and the dwarf the recessive trait. Using capital letters for the dominant trait and lowercase letters for the recessive, he could explain the hereditary relationships this way:

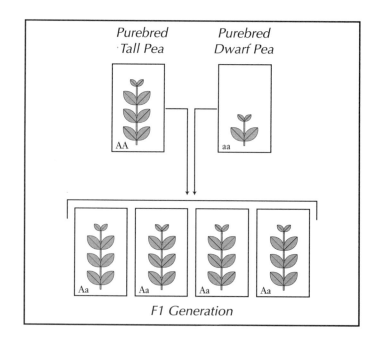

Because tall is dominant, all F1 plants were tall. Now suppose the F1 generation self-fertilized. Both parents of each offspring would have one tall factor (A) and one dwarf (a). The possible combinations for this trait in the F2 generation would be:

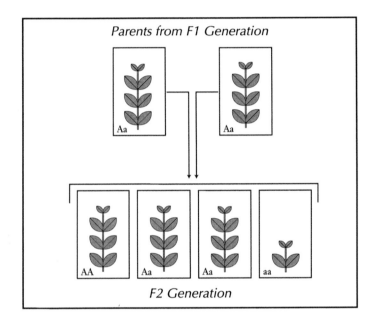

Parents from F1 Generation

F2 Generation

Mendel noted that all combinations containing the dominant tall factor (A) would produce a tall plant. But the combination of two recessive dwarf factors (aa) would produce a dwarf variety. How often would this occur? As illustrated above, Mendel discovered that three out of every four plants in the F2 generation contain a tall factor (A). Therefore, three out of every four of those plants would be tall and only one would be short. Mendel's statistics showed that this was indeed the case.

Mendel concluded, therefore, that hereditary factors occur in pairs. These pairs separate during the formation of sex cells so that the sperm and eggs of the parents contain only one half of each pair. In fertilization, the female sex cell combines with the male sex cell to give the offspring the complete pair of that hereditary factor. Because the pair of factors do not blend, there must be some individual, indivisible unit of heredity. From his experiments, Mendel also concluded that each trait is passed on independently of any other trait.

THE RESULTS

On February 8, 1865, after eight years of careful documentation, Mendel presented his findings to the Natural Science Society in Altbrünn. Because his report was so lengthy, he had to wait until the society's next monthly meeting to complete it.

In many ways, Mendel was far ahead of his time, especially in his use of statistics and ratios. Although the society members understood little of what he was saying, they invited him to publish his report in their journal. Mendel accepted the offer.

Mendel now tried to reach scientific experts who would be more likely to recognize the significance of his work. Unfortunately, the man he confided in the most, German biologist Karl von Nägeli, offered no encouragement. Historians believe that von Nägeli did not accept the findings of this obscure monk because he had his own ideas on how heredity worked. No one else paid any attention to Mendel, either. The most influential book on hybrid experiments that appeared after Mendel's report failed even to mention his work. The monk never found either a financial backer or a major publisher interested in his findings.

Being a member of the religious establishment at a time when religion and science seriously opposed one another may also have hampered Mendel, because his work suggested a number of things that might have infuriated his superiors at the monastery. For example, the exact ratio of dominant to recessive traits plainly suggested that the makeup of offspring was a matter of chance, and determined by mathematical odds. This would have met a solid wall of resistance from those who believed that God specifically fashioned each individual for a purpose.

Furthermore, the church would not have been happy about one of their monks bolstering Darwin's claims. By showing that new variations would not be blended away into the general population, Mendel's findings about dominant and recessive factors resolved a major problem in Darwin's theory of natural selection. Instead, these variations could survive intact to produce the favorable adaptations that Darwin had talked about.

Whether out of frustration with fellow scientists who did not understand his work or because of being intimidated by his superiors, the shy, humble Mendel made no further effort to lecture or publish. Of the 40 copies of his paper, only a few were ever found.

Modestly referring to his time-consuming experiments with more than 30,000 plants as "an isolated experiment," Mendel attempted a few more experiments but ran into a series of problems. An attack of insects called pea weevils devastated his inventory and forced him to abandon further experiments with peas. His choice of a replacement plant (suggested by von Nägeli) called *hieracium*, or hawkweed, was futile. *Hieracium* had a far more complicated reproductive system than he realized, and Mendel's mixed results made him wonder if peas were just a quirky exception in nature rather than the rule.

Elected abbot of his monastery in 1868, Mendel abandoned all experimental work three years later. Instead, he confined himself to his administrative duties and was active and well liked in the community. He died in 1884 without ever gaining the slightest recognition for his work on heredity.

During the latter part of the nineteenth century, scientists also exposed two of the most durable falsehoods regarding the secrets of life. In France, Louis Pasteur performed an experiment that destroyed any argument in favor of the spontaneous generation of life. And in Germany, August

Louis Pasteur (1822-1895) devised an experiment that not only proved the existence of microbes in the air, but also disproved the idea that living things, such as maggots in meat or bacteria, could simply appear. This was called the theory of spontaneous generation and many scientists had accepted its validity until Pasteur's experiment.

Weismann built a strong case for what he called the "germ plasm"—the genetic information contained in a sperm and an egg. He argued that the germ plasm is the only way of passing characteristics on to another generation. Weismann saw no mechanism whereby parents could pass on the traits they had acquired during their lifetimes.

By the dawn of the twentieth century, biologists interested in following up on these discoveries were looking into past experiments on heredity. Working independently in different countries, three of them—Hugo De Vries, Carl Erich Correns, and

The work of biologist August Weismann (1834-1914) supported evolution, and Charles Darwin wrote a preface to one of Weismann's books. It was Mendel's pea experiments, however, that began to explain the facts of variation among individuals.

Erich Tschermak von Seysenegg—found Mendel's published report. Each was thunderstruck. The most valuable experiment in the history of heredity had been buried for more than 30 years! In 1865, Mendel had offered proof that the carriers of hereditary traits were indivisible, individual units and were, therefore, probably in particle form. He had shown that units of heredity are separate in the parents and combine independently of other factors to form a new being that is similar, yet not identical, to the parents.

The rediscovery of Mendel's work, combined with the recent discoveries of microbiology, opened up a new area of scientific study. In 1909, Danish botanist W. L. Johannsen named Mendel's unit of heredity the gene, and the study of heredity became known as genetics. Scientists now set off to build on Mendel's work and to uncover more secrets about the formation of new life.

CHAPTER THREE

Thomas Hunt Morgan and the Chromosome

Once scientists decided that genes, or individual particles of heredity, did exist, the next step was to find out where those genes were located. As scientists developed methods of peering inside the microscopic cells that make up every living thing, they discovered some fascinating structures. In 1879, German researcher Walther Flemming used a special dye to expose thread-like pieces of material within the cell nucleus.

Over the next few years, scientists learned more about these threads of material. They discovered that while the number of these structures varied in different types of animals, each individual within a species had the same number. They found that the sex cells of each organism had only half as many of these threads as the organism's other cells and that, just before a cell divides, the threads line up. Flemming named this process of cell division mitosis, from the Greek word for thread.

Born the year after Gregor Mendel presented his paper suggesting paired genes, Thomas Hunt Morgan (1866-1945) would later prove these genes were carried by chromosomes.

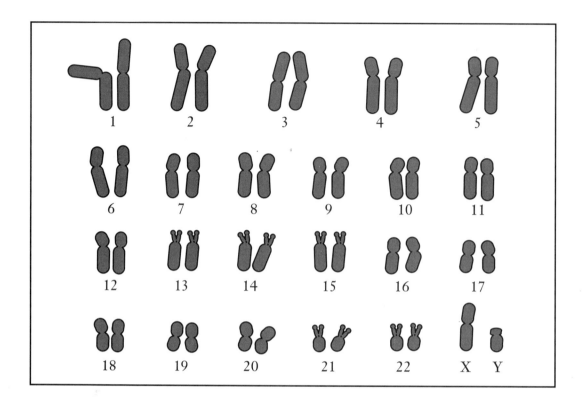

The 23 pairs of human chromosomes carry all the genetic information needed to create a new person.

In 1902, American researcher Walter Sutton observed that these thread-like structures—which had recently been named *chromosomes*, meaning "colored bodies"—existed in pairs in a normal cell. He suggested that these chromosomes carried the paired genes that Gregor Mendel had proposed.

Thomas Hunt Morgan, one of the foremost zoologists in the United States at the time, doubted the theories of Mendel and Sutton. But Morgan was open-minded enough to examine the evidence for himself. He not only changed his mind but also did more than any other person to win acceptance for the chromosome theory of heredity.

Thomas was born on September 25, 1866, to Charlton Hunt Morgan and Ellen Key Howard Morgan. The Morgans were one of the most prominent families in Lexington, Kentucky. One of Thomas's great-grandfathers had set up a small store and a trading route and was thought to be Kentucky's first millionaire. John Morgan, Thomas's uncle, had won fame during the U.S. Civil War as the leader of a Confederate cavalry unit known as Morgan's Raiders. Before the Civil War, Thomas's father, a government diplomat, had often been stationed in Europe. After the war ended in 1865, he had worked as clerk of a U.S. Senate committee and later was appointed a tobacco weigher.

Like Charles Darwin, young Thomas Morgan loved wandering the hills and river valleys, exploring nature. He often led groups of friends on trips to collect butterflies, birds, and fossils.

Thomas attended public school in Lexington and, at age 16, enrolled at the State College of Kentucky (now the University of Kentucky), also in Lexington. While in college, Morgan showed unusual promise in his favorite subject of biology and conducted field studies for the U.S. Geological Survey. After graduating at the top of his class in 1886, he enrolled at Johns Hopkins University in Baltimore, Maryland, where he earned a Ph.D. degree in 1890 for his study of sea spiders.

In 1891, Morgan signed on as professor of biology at Bryn Mawr College in Pennsylvania. One of his students was Lilian Sampson. Years later, in 1904, the two married. They would eventually have

Another of Thomas Hunt Morgan's famous relatives was his great-grandfather, Francis Scott Key. In 1814, Key wrote the words to "The Star-Spangled Banner," the national anthem of the United States.

One of the first cytologists, or scientists who study cells, Walther Flemming (1843-1905) observed cells dividing and named the process mitosis.

Isabel Morgan, one of Thomas Hunt Morgan's four children, was a scientist like her father. In 1948 Morgan developed a polio vaccine that worked in monkeys. This was one of several important discoveries that eventually led to a vaccine against polio for humans.

four children. For many years, Lilian Morgan devoted herself to their children and to shielding her husband from the details of daily life so he could concentrate on his scientific work. At the time of their marriage, women had little chance for advancement in the scientific world. But Lilian Morgan, who had been a talented cell biologist, returned to the laboratory after their children started school and contributed to her husband's work.

Ever since Thomas's early boyhood explorations, he had been fascinated by the process of animals developing from eggs into adults. He continued his research while teaching and, in 1897, he wrote his first book, *The Development of the Frog's Egg: An Introduction to Experimental Embryology.*

In 1904, Morgan began to teach zoology, the study of animals, at Columbia University in New York City. The recent rediscovery of Mendel's research excited him, and he began to delve deeper into the growing field of genetics.

Morgan was far from the stereotypical picture of the serious, obsessed, and socially inept scientist. He was open and friendly with everyone, including his students. Gentle and easygoing, he loved to laugh and enjoyed listening to music, studying the arts, and romping with his children. Unlike the loner Mendel, Morgan surrounded himself with other scientists. At Columbia and later at the California Institute of Technology, he put together a genetic research team that was widely admired in the scientific world for its enthusiasm and spirit of cooperation.

Also unlike Mendel, Morgan was far from meticulous in his habits. He ran an informal laboratory, and visitors were sometimes shocked and disgusted by the lack of cleanliness. Morgan also cared little about his personal appearance and often left his black hair and beard uncombed. He sometimes forgot to put on a belt and then would use a rope to hold up his pants. Once when his shirt collar ripped before an important lecture, he simply had an assistant tape it back together.

For all his informality, Morgan kept himself closely focused on his goals. During the day, he worked in his laboratory. He socialized or exercised in the late afternoon and evening, and he wrote at night. He never sat without pulling out something to read. One associate commented, "I never saw a person who wasted so little time."

As a firm believer in the experimental approach to science, Morgan had little patience with scientists who fashioned detailed theories based only on their observations of the world around them. Instead, he wanted to see data derived from experiments. As a result, Walter Sutton's theory that Mendel's hereditary factors were located on chromosomes did not impress him. For several years, Morgan was perhaps the most vocal critic of Sutton's theory because it was unproven. Ironically, he would eventually provide much of that needed proof. Even more ironically, only several weeks before one of his articles criticizing the chromosome theory appeared in print, he would become convinced of the role that chromosomes play.

THE BREAKTHROUGH

In 1908, Morgan began conducting his own genetic experiments. Like Mendel, he realized that choosing the right subject was crucial. For his experiments, he selected not a plant but an insect with the scientific name of *Drosophila melanogaster*. These insects are commonly known as fruit flies because they feed on overripe fruit.

Morgan chose *Drosophila* because of their rapid life cycle. The flies began breeding at four days. This meant that Morgan could produce a new generation in little more than a week. Because fruit flies also laid hundreds of eggs at one time, Morgan could collect reams of data within a very short period. Moreover, the flies displayed a clear genetic variation so that Morgan could easily follow traits as they were passed along to new generations. They were small, only one-eighth of an inch long, and could be stored in a small space. *Drosophila* also had only eight chromosomes (four pairs). If chromosomes turned out to be important to genetics in any way, having so few to study would simplify matters considerably.

Morgan patterned his experiments after Mendel's, choosing a few easily identifiable traits that occurred in pairs. Within a short time, he and his assistants were tabulating the traits of hundreds of thousands of fruit flies.

During the course of his studies, Morgan found a number of cases in which a trait seemed to come out of nowhere. Using a term introduced by

Dutch botanist Hugo De Vries, Morgan referred to these as mutations. Their sudden appearance seemed to support Darwin's ideas. Here were new traits suddenly materializing in the genes, ready to be passed on to future generations if they provided a survival advantage. Or, if the traits proved to be a disadvantage, they were slated for extinction.

The most intriguing mutation involved eye color. All members of a strain of *Drosophila* had red eyes until a white-eyed fruit fly suddenly appeared among them. Morgan mated a red-eyed fly with a white-eyed one and found the offspring (F1) to be red eyed. So the gene for red eyes must be dominant.

Morgan then mated F1 offspring. According to Mendel's experiments, they should produce three red-eyed flies for every white-eyed fly. When he tabulated the results of the large pool of fruit flies that he had collected over several months, Morgan counted 3,470 red-eyed and 782 white-eyed flies. While that was close to the expected ratio of 1 in 4, the really curious feature of the experiment was that not one female had white eyes! Among the males, 1,011 had red eyes and 782 had white eyes.

Morgan could see only one answer to this mystery. All genetic traits were not passed on independently, as Mendel had supposed. Instead, some genetic traits must be linked. The genetic carrier for white eyes was connected to the carrier for the male sex. Scientists had already detected a difference in one of the chromosomes of males and females. If the white-eyed trait was linked to sex, it must also be physically linked to the chromosome.

mutation: a sudden structural change in a gene or chromosome of an organism

Morgan's laboratory at Columbia was affectionately known as "the fly room." This small room, 16 by 23 feet in size, was on the top floor of the biology building. Eight desks were crowded into this room and Morgan had a small office off to one side. The flies were grown in empty milk bottles borrowed from the university's lunch room. The mashed bananas on which the fruit flies thrived also fed a healthy population of cockroaches.

As Morgan and his associates continued their studies, they found more traits that appeared to be linked, although not all of them were linked according to sex. In fact, Morgan eventually discovered four distinct groups of traits that appeared to be inherited together, which corresponded exactly with the number of pairs of chromosomes that *Drosophila* had. Furthermore, Morgan found that one of the four linkage groups had fewer characteristics than the other three. That fact correlated neatly with the fact that one of the *Drosophila* chromosomes was smaller than the other three.

Now the evidence that chromosomes carried genes was overwhelming. Chromosomes, first of all, occurred in pairs within cells. In the creation of

The many black dots Morgan is studying are fruit flies. Behind him, large diagrams of the flies are mounted on the wall.

sex cells, the chromosomes divided so that each sex cell had only one half of each pair. During fertilization, the two groups of chromosomes, one group from the sperm and one group from the egg, joined together to provide a complete set. In each set, one chromosome was different for males and females. Since other inherited traits were linked to gender, they must be also be linked to chromosomes. The *Drosophila* traits passed from generation to generation in four linked groups.

Just as Mendel's conclusion that traits were inherited separately was not accurate, so was Morgan's initial conclusion that traits were permanently fixed to a spot on a chromosome. In sifting through their hundreds of thousands of fruit fly cases, Morgan's team found a few instances where the normal linkage between traits did not occur. For example, fruit flies with black bodies normally had short wings, while those with gray bodies had long wings. Although the genes for those traits appeared to be genetically linked, Morgan came across an occasional black-bodied fly with long wings. Such a finding seemed to scramble his entire theory of linked genes.

When Morgan tried to find an answer to this exception to the linkage rule, he noticed a difference in how firmly certain traits were linked. Some traits occurred together all of the time. Others, however, split apart some of the time, and some almost never split. He observed that one chromosome could get temporarily tangled up with another chromosome during cell division. When this

With self-deprecating humor, Thomas Morgan once said, "I am a professor of experimental zoology and I have three sorts of experiments: fool experiments, damned fool experiments, and those that are still worse."

happened, he theorized that the long chain of the chromosome could break at any point along its length. The broken pieces could then recombine, but they might not re-attach to their original chromosome. When that occurred, normally linked traits could be on separate chromosomes. Morgan reasoned that those traits located far apart on the chain would tend to become separated more often than those that were located close together.

Using this system, Morgan's team measured the frequency with which linked traits separated. From that information, Morgan was able to draw a map of the fruit fly's genes, showing the location of the gene for each trait along the length of the chromosome.

THE RESULT

In 1915, Morgan and his team published *The Mechanisms of Mendelian Heredity*, an examination of the role of chromosomes in heredity. Some scientists resisted the ideas that Morgan proposed. Gradually, however, the weight of Morgan's experimental data overwhelmed his opponents.

The Theory of the Gene, which Thomas Morgan published in 1926, provided a solid basis for understanding how genes transmitted traits from parents to their offspring. That year, Morgan came across convincing proof of his gene theory when he visited the laboratory of Barbara McClintock at the Cornell College of Agriculture in Ithaca, New York.

McClintock was a brilliant geneticist who specialized in the study of multicolor maize, or Indian corn. Using new staining techniques that under a microscope revealed tiny knobs and other physical differences, she learned to identify each of the corn's 10 chromosomes by sight. She then studied certain visible genetic traits.

She and her assistant, Harriet Creighton, traced the characteristics of waxy, purple kernels to the ninth chromosome in a particular strain of corn. This chromosome was easy to distinguish because of a knob on one end and a long, narrow tip on the other. McClintock's studies convinced her that the gene for waxiness was located near the narrow tip and the gene for purple kernels near the knob.

McClintock fertilized this strain of corn with a different strain. At harvest time, some ears were

Barbara McClintock (1902-1992) holds the multicolored maize, or Indian corn, that she grew to demonstrate that genes can change places on chromosomes.

both waxy and purple, some were neither waxy nor purple, and some were one but not the other. She saw that these features corresponded to changes in the knobs and the tips of the chromosome. (In some cases, the knobs and tips actually changed places.) Her experiments demonstrated that chromosomes carried genes for physical traits and could create variety by exchanging parts. This helped to account for the never-ending introduction of variety into species that Darwin had assumed but had not been able to explain.

McClintock was reluctant to publish her findings until she gained further proof. Thomas Morgan was so excited by her discovery that he wrote to a scientific journal, saying an important article would arrive in two weeks. He then persuaded McClintock to write the article.

McClintock took this explanation of variety one step further. Her discovery of genetic mutations caused by X rays convinced her that genes were not as stable as Morgan had thought. McClintock doubted whether genes really did occupy permanent spots on the chromosome that could be shifted only if the chromosome broke.

Working alone at the Cold Spring Harbor Research Center, on New York's Long Island, McClintock studied multicolored maize whose chromosomes had been injured. She found strange patches of colors occurring in pairs and deduced that some change must be taking place as the plants developed. Then she noted that when chromosomes repeatedly broke and rejoined, they gained, as well as lost, information.

After six years of research, McClintock knew why. Her microscopic studies showed that not only could genes move around to different positions on a chromosome, but they could also have some type of control switches that could turn them on and off at different times.

Morgan had died in 1945, so he could not benefit from this discovery. In fact, for quite some time, few scientists benefited from McClintock's discovery; because her finding was so unexpected, she was either laughed at or ignored. Not until the 1970s did geneticists find that these same "on-off switches" and changes of position occurred in all forms of life. This confirmation won McClintock a long-deserved Nobel Prize in physiology in 1983.

The phenomenon of "jumping genes," as scientists playfully called them, completed the story of the chromosome begun by Thomas Hunt Morgan. As developed by Morgan, the chromosome theory allowed scientists to pinpoint the actual location of genes on chromosomes and to explain how genes passed from one generation to another. With refinements by McClintock, this theory also shed light on the source of the incredible variety of life. It helped to explain such mysteries as how new strains of plants and bacteria could evolve; how human cells could continually develop new antibodies in the blood to protect against new forms of disease; and how species developed new features, some of which gave them a survival advantage over others of their species.

Morgan was proud of his role in shaping the science of genetics. "The whole subject of human heredity in the past has been so vague and tainted by mystery and superstition that a scientific understanding of the subject is an achievement of the first order," he said. In recognition of his achievement, Thomas Hunt Morgan was awarded the Nobel Prize in medicine and physiology in 1933.

Thomas Hunt Morgan signs autographs for neighbor children on October 23, 1933, the day it was announced that he had won the Nobel Prize for his work in genetics. He insisted that photographers include the children. Morgan split the prize money between his children and the children of his colleagues, Calvin Bridges and A. H. Sturtevant. While generous with his own money—he funded several anonymous scholarships, for example— Morgan took pride in being frugal with research funds.

CHAPTER FOUR

Oswald Avery
and the Transforming Principle

In 1869, Johann Friedrich Miescher (MEE-sher), a 25-year-old Swiss chemist, found that science sometimes required a strong stomach as well as a keen mind. In order to obtain a sample of human lymph cells, he collected used bandages from a local hospital. Miescher separated the pus from the bandages and then tried to analyze it to see what its cells were made of. He was not interested in the central part, or nucleus, of the cell. Instead, he focused on the fluid found within the cell wall that was known as the cytoplasm.

In trying to isolate large particles from the cytoplasm, Miescher ran into one problem after another. One experiment yielded a strange substance that was unlike any protein he had previously encountered. The young scientist noticed that the substance appeared only when he added a weak alkaline solution. Since alkaline solution was known to cause nuclei to burst open, Miescher realized that he

Oswald Avery (1877-1955) was a quiet man who made a big noise in the world of genetics when he discovered that DNA carried genetic information.

71

The average animal cell is about 0.0004 inches across. Cells vary in shape according to the job they perform. Most, but not all, have a control center called the nucleus. The nucleus contains the cell's genetic information and controls the cell's actions. (Nuclei is the plural of nucleus.)

had probably failed to separate the cytoplasm from the nucleus and that the material was from the nucleus itself.

Because he was having little luck with his cytoplasmic experiments, Miescher shifted his focus to the nucleus and the strange substance. To the nucleus, he added an enzyme (a substance that speeds up certain chemical reactions) called pepsin. Although pepsin caused the proteins to break into smaller pieces, it had no effect on this material. This was confirmation that the substance was not a protein. When Miescher analyzed this material further, he was amazed to discover that it contained nitrogen and phosphorus, a mineral common in rocks but rare in living things. Since he had discovered the unknown substance in the cell nucleus, he called it nuclein.

Miescher's supervisor, a distinguished German scientist named Ernst Hoppe-Seyler, cast a suspicious eye on Miescher's report. The results were so different from what Hoppe-Seyler had expected that he thought the inexperienced Miescher must have made a mistake. Hoppe-Seyler himself had discovered lecithin, the only natural substance known at the time to contain both nitrogen and phosphorus. Only after the older man ran all the younger scientist's experiments again and obtained the same results were Miescher's findings published.

Neither Miescher nor Hoppe-Seyler, however, grasped the significance of their discovery. In fact, more than 70 years would pass before scientists recognized the nuclein as the basic building block of all life. That discovery came about through the efforts of a tiny, bald medical researcher named Oswald Avery, and it caught the scientific world completely by surprise.

Oswald Theodore Avery was born in Halifax, a city in the Canadian province of Nova Scotia, on October 21, 1877. His father, a clergyman, moved the family to the United States a few years later.

Oswald was small for his age throughout his childhood and would weigh less than 100 pounds as an adult. After he graduated from high school, he attended Colgate University and then moved on to Columbia University in New York City, where he studied to become a physician. Avery was a good student, but he was so shy and quiet by nature that his classmates did not think of him as particularly capable.

After receiving his medical degree in 1904, Avery served as an assistant to a physician for several months. During that time, like Charles Darwin, he decided that he was really not cut out for a career of treating patients. He was far more interested in medical research, particularly research dealing with the nature of infectious diseases.

In 1906, Avery accepted a job at a medical laboratory in Brooklyn, New York. After he had worked there for nearly seven years, he published a paper about tuberculosis, which caught the attention of the director of the Rockefeller Institute Hospital in New York City. (The Rockefeller Institute was one of the few facilities in the United States that funded medical research on a large scale.)

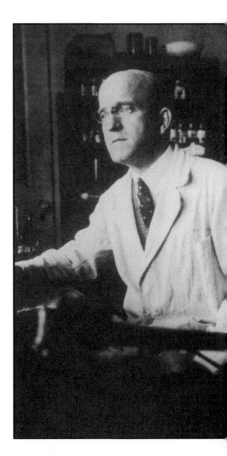

Oswald Avery in his laboratory at the Rockefeller Institute, which was later renamed Rockefeller University.

pneumonia: a disease that is caused by viruses, bacteria, or other microorganisms and characterized by the swelling of the lungs. From the Greek word *pneuma,* which means air or breath.

Avery joined the institute as a bacteriologist in 1913 and was assigned to study the deadliest substance on earth—the microorganisms that caused pneumonia. At that time, pneumonia was by far the leading cause of death in the world.

Coming to the institute would be the last career move of Avery's life. He would spend the next 35 years there, earning high praise from his colleagues. In fact, he explained concepts and techniques so well that his laboratory mates nicknamed him "Professor," which they soon shortened to "Fess."

Avery thrived so on his work that he seemed to have little need for anything else in his life. He rarely socialized and never married. He shared a home with another scientist, Alphonse Dochez, and the two of them seldom left their work at the office. Instead, they would often discuss projects far into the night.

At the beginning of his most important work in the early 1930s, Avery struggled with a severe case of Graves' disease, which affects the thyroid gland. As a result, he suffered from depression, weight loss, and developed a tremor in his hands that hampered his ability to do some of the delicate work involved in microbiology. For a man dedicated to precision and thoroughness, this loss of control in the laboratory was devastating. In the mid-1930s, he finally underwent thyroid surgery that eased his condition considerably. It was at this time that Avery found himself drawn into a project that had developed quite unexpectedly.

THE BREAKTHROUGH

In 1928, Frederick Griffith Jr., a medical officer at the Ministry of Health in London, made an interesting discovery. While looking for a way to help fight pneumonia, Griffith performed a study on two strains of a type of bacteria (known as pneumococci) that caused the killer disease. One of the two strains was deadly to humans; the other was harmless. Griffith found that the only physical difference between the two was that a smooth-coated capsule made of sugar surrounded the deadly strain while the harmless strain had a rough-coated surface with no such capsule. The white blood cells in the human body that were responsible for destroying bacterial invaders had trouble penetrating the smooth coat.

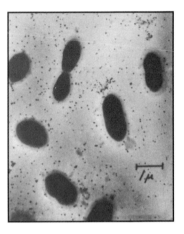

Oswald Avery experimented on a type of pneumococci similar to these.

While experimenting with these two strains, Griffith stumbled across an event so bizarre that it seemed like something out of science fiction. He killed some of the deadly, smooth-coated pneumococci (called pneumococcus S, for smooth) with heat and then injected them into mice. As expected, the mice that received the dead bacteria did not develop pneumonia. Next he injected the live, rough-coated bacteria (called pneumococcus R, for rough) into the mice. Again, as expected, nothing happened. But when Griffith injected the killed pneumococcus S along with the harmless live pneumococcus R, the mice caught a raging case of pneumonia and died!

Either the dead pneumococci had come back to life or something had happened to transform the harmless bacteria. Griffith discovered that the rough-coated, harmless bacteria had acquired the smooth coat of the deadly strain and had become infectious. Even more incredible, the bacteria that had acquired this smooth coat were able to pass this trait on to the next generation. The change, therefore, was both permanent and inheritable.

What had caused this change? Griffith speculated that the benign strain had become harmless because it had somehow lost the gene for producing the smooth coat. Perhaps something in the killed bacteria was able to restore to the live bacteria the ability to make a smooth, deadly coat. Since the substance came from dead bacteria, the gene itself was not alive. In that case, the instructions contained in the genes must have somehow been transmitted by chemical compounds.

Oswald Avery greeted Griffith's report on these findings with some skepticism. After all, Avery himself had been working with pneumococci for more than a decade and had never come across any situation similar to what Griffith had described. When he asked his assistants to duplicate Griffith's work, they not only proved Griffith correct but were also able to change harmless, rough-coated bacteria into dangerous smooth-coated bacteria inside laboratory glassware.

Avery's team then took the process a step further. They alternately froze and thawed smooth-coated pneumococci to break open the cells. Using a centrifuge, a machine that spins objects at high speeds, they separated the cell fluids from the broken cell debris. When they added this fluid to

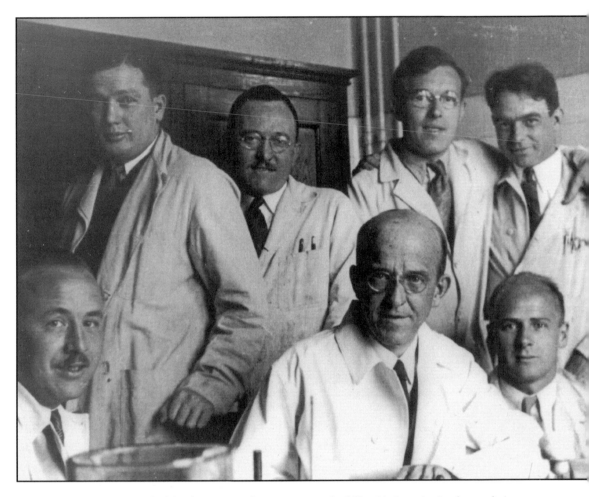

Oswald Avery surrounded by his research team at Rockefeller University in the early 1930s. From left to right are Thomas Francis Jr., Ed Terrell, Kenneth Goodner, Rene Dubos, Frank Babers, and Walther Goebel.

amino acid: an organic compound that is the basic building block from which proteins are synthesized in the cell

atom: the smallest unit of an element, consisting of electrons that surround a central, dense nucleus

protein: any of a group of complex organic molecules that consists of chains of amino acids, and contains substances, such as enzymes, hormones, and antibodies, that regulate body functions

rough-coated pneumococci, the bacteria developed the deadly, smooth coats.

At this point, Avery became personally interested in what he nicknamed "the sugar-coated microbes." (Microbe is another name for a microscopic organism.) Although many scientists greeted Griffith's report with curiosity and amazement, only Avery realized that the extract from the dead bacteria held the key to a tremendous breakthrough in genetics. If he could purify the substances in the fluid and individually test them, he might be able to discover the chemical responsible for passing along inherited traits.

Virtually every scientist who considered the matter thought the substance that carried hereditary information had to be a protein. They believed that enormous molecules would be required to account for the almost limitless amount of variety carried by the genes. If a gene could express itself in a million different forms, then the substance that carried it had to be able to do the same. Proteins were the logical choice because they were extremely complex molecules consisting of chains of smaller molecules called amino acids. Amino acids were made up of many atoms, so a single protein molecule might contain hundreds of thousands of atoms.

Avery and his team set out on a careful, step-by-step plan to purify what they called the "transforming principle" in the pneumococci. They grew large amounts of deadly bacteria in 20-gallon vats of broth made from beef hearts that was heated to the temperature of human blood. They then spun the broth in a centrifuge to separate the pneumo-cocci. Next, they broke open the cells to extract fluid. Using various chemicals and enzymes, they began to extract pure substances from the fluid. Finally, they began mixing these pure substances, one by one, with harmless pneumococci. The process took many years. In May 1942, Avery expressed his frustrations to his brother in a letter in which he wrote that this was "some job—full of headaches and heartbreaks." None of the proteins that he had tried so far had affected the pneumococci bacteria.

Only a nonprotein substance showed promise. When Avery performed a chemical analysis of the one purified substance that could induce a smooth, deadly coat in harmless pneumococci, he found that its ratio of nitrogen to phosphorus was 1.67 to 1. This happened to be the ratio found in a chemical compound known as deoxyribonucleic acid, or DNA for short. Further tests verified that the material was, indeed, DNA.

The presence of DNA in the cell nucleus was certainly no secret. Johann Friedrich Miescher had discovered that substance many decades earlier, naming it nuclein. But virtually no one had suspected that DNA played any role in heredity. Although it was a very large molecule, it seemed far too simple to accommodate the billions of forms that were needed to account for all of the variety in the world. DNA was made up of nothing but phosphate; a sugar known as deoxyribose; and some combination of only four small, nitrogen-based molecules known as

DNA (deoxyribonucleic acid): a nucleic acid that carries the genetic information in the cell; it consists of two long chains of nucleotides that determine individual hereditary characteristics

enzyme: a protein produced in cells that speeds up the rate of biological and chemical reactions

nucleotides. Most scientists believed that the DNA molecule was a long chain of monotonously repeating nucleotide sequences.

The likelihood of DNA being actively involved in heredity seemed so small that Avery thought he must have made a mistake. Perhaps he had not purified this transforming agent as thoroughly as he should. Concerned that a small amount of protein still contaminated his DNA fraction, he added enzymes to his transforming agent, knowing that they would break down (or digest) any protein that was still present. When he introduced the resulting product to harmless pneumococci, the smooth coats appeared.

When the enzyme did not affect the transforming power in the least, Avery finally became convinced that protein was not involved in this process. As a final test, he added a substance known to destroy DNA to his transforming agent. The substance instantly lost its ability to transform rough-coated pneumococci into smooth.

THE RESULT

Ever cautious, Avery did not report his findings until he had put together an airtight case for claiming that DNA was the transforming agent. In February 1944, nearly two years later, he and two colleagues, Maclyn McCarty and Colin MacLeod, published a paper detailing their experiments. Even then, he hedged slightly, admitting that there remained a slight possibility that some undetectable trace impurity had caused the change in the bacteria. He concluded by stating that "the evidence present supports the belief that a nucleic acid of the deoxyribose type is the fundamental unit of the transforming principle of pneumococcus Type III."

Maclyn McCarty (above) and Colin MacLeod

Avery, however, knew that he had opened the door to one of nature's most valuable secrets. After centuries of speculation, scientists could at last put their hands on a pure substance that carried genetic information. "This is something that has long been the dream of geneticists," Avery wrote in a letter. Daring to dream a little, he proposed that scientists might be able to use his discovery to cause hereditary changes artificially.

The enormity of Avery's discovery was not lost on other scientists, who read his report with astonishment. As carefully documented as Avery's report was, the idea that an apparently unimpressive molecule such as DNA was the carrier of heredity was too unbelievable for many of them. Some attacked Avery's work and insisted that only a protein was complex enough to pass along genetic information. There must have been impurities in Avery's sample, these scientists maintained. Until the late 1940s, a few biochemists remained skeptical of the role of DNA.

Even if Avery had been wrong, he still would have played a major role in the science of genetics simply by igniting a bonfire of interest in the subjects of genetics and molecular biology. For aside from Avery and his colleagues, no one else had taken the trouble to pursue the promising lead that Griffith had developed nearly 15 years earlier.

Once Avery's report came out, geneticists could hardly talk about anything else. During 1944 alone, his work was the subject of six major conferences. Many brilliant scientists who had previously shown no interest in genetics eagerly rushed into the field to explore the exciting new possibilities.

Oswald Avery had not been wrong and the weight of his research soon swayed most scientists to support his position. Although he never won a Nobel Prize for his breakthrough, he did witness the conversion of his theory into established fact. Just two years before Avery's death on February 20, 1955, scientists James Watson and Francis Crick explained exactly how DNA accomplished its task of passing along genetic information to new generations.

Avery was the world's first genetic engineer—the first to use DNA to introduce a gene (for a smooth coat) into an organism that did not previously have that gene. His discovery of DNA as the transforming agent altered the landscape of genetic research forever. From that time on, geneticists would make their major advances by thinking small. As researchers began to delve into a realm that the average person could barely comprehend, the study of heredity began to a focus on the mysterious workings and interactions of unseen molecules.

Oswald Avery spent 35 years working in the laboratories at Rockefeller University. He lived long enough to see that his discovery led to further breakthroughs.

James Watson (left) and Francis Crick unraveled the mystery of the chemical nature of DNA. Watson called it the most important discovery in the life sciences of the twentieth century.

CHAPTER FIVE

James Watson and Francis Crick and the Double Helix

After Oswald Avery identified the chemical compound that carried genetic information, the next important question was how a chemical could pass on instructions for building living things. Even more baffling was how DNA—a chemical compound with relatively few parts—could carry the genetic blueprints for every component of every creature that had ever lived.

That question had much more to do with chemistry than with conventional biology. In order to find out how DNA passed along information, researchers had to figure out exactly how the DNA molecules were put together. Prior to Oswald Avery's discovery, few geneticists had a solid enough background in chemistry to know how to approach the problem.

Eventually, three different high-powered teams of researchers recognized the importance of identifying the structure of DNA. In contrast to the gentlemanly days when naturalists Charles Darwin and Alfred Russel Wallace had politely engaged in their "you first, no, you first" etiquette regarding the publication of their papers about natural selection, the search for the structure of DNA spawned a fiercely competitive race among these three groups.

No one was more determined to win that race than a young American biologist named James Watson. Despite the fact that he considered himself lacking in mathematical skills and had little knowledge of nucleic acids, Watson wanted desperately to be a part of that history-making achievement.

James Dewey Watson was born on April 6, 1928, in Chicago, Illinois, to Jean Mitchell Watson and James Dewey Watson Sr. His sharp mind quickly impressed teachers at Horace Mann Grammar School and South Shore High School, and he was a "Quiz Kid" on national radio. Far ahead of his high school class, he accepted a scholarship to the University of Chicago when he was just 15 years old. Particularly interested in birds, Watson earned his college degree in zoology before his twentieth birthday.

While doing postgraduate work at Indiana University, Watson became interested in a group of tiny viruses known as bacteriophages and earned a Ph.D. degree for his studies in this area. At the urging of his academic advisor, in 1951, he traveled to Copenhagen, Denmark, on a research grant to study more about the DNA of these bacteriophages. But Watson found the subject to be both boring and difficult. He once admitted that "it was my hope that the gene might be solved without my learning any chemistry."

While in Europe, Watson met British scientist Maurice Wilkins from King's College in London. When Wilkins spoke about efforts underway in DNA research, Watson realized what the discovery of the structure of DNA would mean to biology. He then became so excited about the possibilities that he abandoned his studies and joined another DNA research group at Cambridge University. There Watson stirred up the interest of Francis Crick in exploring the nature of the genetic molecule.

James Watson in his laboratory at Harvard University in 1962. He came to Harvard in 1956 and studied the role of DNA in protein synthesis there until he became the director of the Cold Spring Harbor Laboratories in 1968.

Francis Harry Compton Crick was born in Northampton, England, on June 8, 1916. His family owned a prosperous boot and shoe factory. Francis attended a private school for boys called Mill Hill on a scholarship, and he moved on to study physics at University College in London. He was working on an advanced degree when World War II broke out. Pushed to the edge of defeat by Adolf Hitler's Nazi forces, the British had to focus all of their efforts on survival. Like many other young people, Crick abandoned his studies and put his inventive mind to work for the British military.

While working on new radar technology and helping to design such ingenious devices as magnetic mines, Crick lost interest in his former studies.

Sidetracked by working for the British military during World War II, Francis Crick returned to college in 1947, at the age of 31.

When he finally returned to college, Crick decided to abandon physics and explore an entirely different scientific subject—molecular biology. In 1947, supported by a government grant, he began working on his Ph.D. degree at Cambridge University.

Some of his colleagues at Cambridge found Crick's manner irritating. He talked loud and fast, often punctuating his words with a booming laugh. When he came up with a new idea, he would practically burst with enthusiasm. These ideas often amounted to nothing, and he would return to his experiments only to rush back a day or two later, excited about another thought.

Watson and Crick seemed an unlikely pair to make a breakthrough in DNA studies. Watson was much younger than many of the world's top researchers. Slender, nervous, and boyish in appearance, he looked even younger than he was. He also tended to be reckless and impatient in his approach to science and once nearly destroyed a laboratory when he got careless with a Bunsen burner, a small gas-flame device. He was equally brash in his disregard for the traditional gentlemen's agreements between scientists. According to etiquette, British research involving DNA was Maurice Wilkins's domain because he had been the first in the country to start the project. Watson ignored the tradition that others should leave Wilkins alone and instead pursue their own projects.

Crick had more training than Watson in the biochemical part of DNA research, but he could be just as inept in a laboratory. He twice flooded the lab

"Jim and I hit it off immediately, partly because our interests were astonishingly similar and partly, I suspect, because a certain youthful arrogance, a ruthlessness, and an impatience with sloppy thinking came naturally to both of us."—Francis Crick

"From my first day in the lab I knew I would not leave Cambridge for a long time. Departing would be idiocy, for I had immediately discovered the fun of talking to Francis Crick. Finding someone in [the] lab who knew that DNA was more important than proteins was real luck."—James Watson

at Cambridge because he forgot to fasten a simple rubber tube during an experiment. Despite his inventive mind, his career did not seem to be going anywhere. In 1951, he and his second wife, Odile, were living in a tiny apartment. Already in his mid-30s, Crick still was far from meeting the requirements for his Ph.D degree.

When Watson arrived at Crick's laboratory, he provided both friendship and a strong focus for Crick's flashes of brilliance. Before long, Crick caught Watson's enthusiasm for DNA. The two bounced ideas off each other and debated various approaches. What would DNA look like? What construction of the molecule could account for both the available facts about DNA and for everything known about heredity?

THE BREAKTHROUGH

When Watson and Crick began tackling the problem of DNA in 1951, scientists already understood its basic components. They knew that DNA was a tremendously long molecule composed of smaller units called nucleotides. Each of the nucleotides consisted of sugar, phosphate, and one of four nitrogen-based compounds: adenine, guanine, thymine, and cytosine. Scientists agreed that the nitrogen-based compounds were the active ingredients in heredity. The sugar and phosphate probably functioned as a frame or backbone that supported the long molecule and held it together.

nucleotide: a compound consisting of a base, a sugar, and a phosphate group; DNA and RNA are made up of nucleotides

helix: a spiral form or structure. Helices is the plural of helix.

Watson and Crick decided to use a technique developed by American chemist Linus Pauling. In working out the structure of molecules, Pauling manipulated building blocks that looked much like children's toys. Each piece was shaped according to the known characteristics of the atom it represented. To deduce the structure of the molecule, Pauling simply observed the most logical way for certain pieces to fit together.

Unfortunately, the components of a DNA molecule could fit together in a number of ways. Again, Watson and Crick followed Pauling's lead. Earlier that year, Pauling had demonstrated that chains of amino acids (the basic units of protein molecules) could arrange themselves in the form of a twisting spiral, known as a helix. Watson and Crick wondered if chains of nucleotides that made up DNA also twisted into a helix. They attempted to twist their DNA building blocks into such a shape, but nothing they tried seemed to work.

At this point, Watson and Crick turned to a technique known as X-ray crystallography which produced patterns as X rays bounced off sections of molecules. They recorded the photographic image that these collisions caused. Watson and Crick hoped that the images provided from this series of hits would outline the general form of the molecule.

Neither Watson nor Crick, however, had any expertise in X-ray crystallography. Instead, they had to depend on data developed by their rival, Maurice Wilkins, and his colleague, Rosalind Franklin. In late 1951, Watson heard Franklin give a lecture on her subject. Neglecting to take any notes, Watson tried to give Crick a summary of what she had said. Based on Watson's secondhand report, the two men excitedly put together a model that had the phosphate component at the center of the molecule. Everything seemed to go together well, so they eagerly called in Franklin and Wilkins to show what they had accomplished.

Watson, however, had misunderstood Franklin. She immediately saw that, based on her research, their model was absurd. Irritated at their carelessness, she was none too gentle in showing them their mistake. The cold shock of embarrassment cooled Watson's and Crick's enthusiasm for DNA research. During late 1951 and early 1952, they quietly worked on projects unrelated to their DNA search. But the urge to solve the DNA mystery returned, and by mid-1952, the two scientists were as deeply

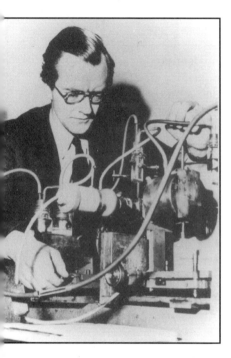

Born in New Zealand and raised in England, Maurice Wilkins helped the United States develop the atomic bomb during World War II. Disgusted by the use of the bomb, he afterwards turned his attention to biological questions.

engrossed as ever in unlocking the key to the structure of DNA.

In December of that year, Watson and Crick received crushing news from Linus Pauling's son, who happened to be studying at the Cambridge laboratories. Pauling had written to his son that he had worked out the structure of DNA, and he had included a report on his work. Heartbroken that Pauling had beaten them, Watson and Crick read the report that Pauling's son showed them. To their astonishment, they discovered an error in Pauling's work. He had not solved the problem after all!

Watson knew that a scientist as great as Pauling would soon find his error and would be more determined than ever to set things right. He guessed that he had about six weeks before Pauling would succeed. The urgency spurred Crick and Watson into one last great effort to solve the puzzle.

Two intense rivalries aided the pair in their work. First, Maurice Wilkins also feared that Pauling would soon make the breakthrough. Having been beaten out by Pauling on other discoveries, Wilkins was eager to see a British group come out on top. Thus, he was willing to cooperate with Watson and Crick. He told them that according to his studies, the diameter of a strand of DNA was thicker than would be expected on a single, long chain.

Furthermore, Wilkins knew that Rosalind Franklin had recently made the clearest X-ray photographs of DNA yet. She was extremely cautious, however, and did not want others to see her data until she had put them together conclusively.

In 1939, Linus Pauling (1901-1994) published The Nature of the Chemical Bond, *one of the most influential chemistry texts of the twentieth century.*

At this point, Watson and Crick benefited from a turf war between Franklin and Wilkins. Wilkins assumed that he was in charge of all phases of DNA research at King's College. He expected Franklin to serve as his assistant and provide him with data at his request. Franklin, however, believed that she was hired to undertake the X-ray crystallography of DNA as her own separate project. Their misunderstanding led to a battle of wills and such hard feelings that Wilkins, without his colleague's permission, showed Watson a print of Franklin's latest photograph.

According to Watson, "The instant I saw this picture my mouth fell open and my pulse began to race." The photograph clearly indicated that the molecule was in the shape of a helix.

Watson and Crick went back to their models and again tried to create a DNA molecule that fit the evidence. But as they continued to arrange the nitrogen bases around an inside backbone of sugar and phosphate, they realized that such a model would not work.

Suddenly, Crick noticed a crucial clue that had been right under their noses for several years. In the late 1940s, Erwin Chargaff had measured the amounts of the nitrogen bases in pieces of DNA. He had discovered that in every case, the amount of adenine was about equal to the amount of thymine and the concentration of cytosine was about equal to the concentration of guanine.

Crick realized that adenine and thymine could be joined together in a hydrogen bond, as could cytosine and guanine. If there were always equal amounts of adenine and thymine, the logical explanation was that they always appeared together in the molecule.

With the discovery that nitrogen bases occurred in pairs, that DNA took the form of a helix, and that the molecule was thicker than a single strand, the pieces quickly fell together. Watson and Crick concluded that the DNA molecule took the coiled, twisted form of a double helix. If the nitrogen bases bonded together, as Erwin Chargaff's study suggested, then they had to be in the center of the two sugar-phosphate supports, just as the steps of a ladder have to be in the center between the two sides of the ladder.

By March 7, 1953, Watson and Crick had created a complete model of DNA that appeared to account for all of the available facts. Then they began to prepare a report that would announce their findings to the world.

"The final version was ready to be typed on the last weekend of March. Our Cavendish typist was not on hand, and the brief job was given to my sister. There was no problem persuading her to spend a Saturday afternoon this way, for we told her that she was participating in perhaps the most famous event in biology since Darwin's book. Francis and I stood over her as she typed the nine-hundred-word article that began, 'We wish to suggest a structure for the salt of deoxyribose nucleic acid (DNA). This structure has novel features which are of considerable biological interest.'"—James Watson

RESULT

Watson and Crick just beat their competition to the finish line. Two days after the pair had solved the DNA puzzle, Wilkins wrote and told them that, based on his latest findings, he expected to uncover the structure of DNA very shortly. When Watson and Crick told Wilkins that they had already done the job, Wilkins was gracious about losing the race. Although he called the pair "rogues" (or thieves), he did so good-naturedly while offering them his congratulations.

On April 25, 1953, Watson and Crick officially claimed credit for their discovery by publishing a 900-word report in *Nature* magazine. Their model was so logical and persuasive that scientists immediately accepted their findings as valid. As Watson had anticipated, he and Crick earned instant recognition as two of the most renowned scientists of all time. In 1962, they shared a Nobel Prize with Maurice Wilkins for their DNA work.

The manner by which Watson and Crick had won their race generated controversy over the years. In particular, the tragic fate of Rosalind Franklin cast a cloud over the breakthrough. Franklin died of cancer at the age of 38 without ever being fully aware of the role she had played in the discovery of DNA's structure. According to her working notes, she had deduced that the shape of DNA was a double helix before Watson had, and she had been the one who had developed the data that eventually put Watson on the right track. Had Wilkins not secretly given

*The Watson and Crick
double helix model*

her findings to Watson, Franklin might well have been the first to publish proof that DNA took the shape of a double helix. In a more cooperative atmosphere, she might have at least shared in the publication of the findings with the others.

What is not debatable, however, is the importance of the double helix model that Watson and Crick worked out. It offered a simple, yet fascinating, explanation of how living things can create new life so similar to themselves and yet allow for almost infinite variation.

The letters A, T, G, and C represent the four nitrogen bases in the structure of DNA. Adenine always combines with thymine and cytosine always bonds with guanine.

According to the Watson-Crick model, the nitrogen bases are paired in the center of the spiraling molecule, held together by a weak hydrogen bond. This can break apart so easily that the entire strand of DNA can split in two, just as a zipper splits apart two sides of a jacket. When a chromosome in a cell is ready to divide and needs a duplicate chromosome to pass on, an enzyme cuts through, or unzips, all of the hydrogen bonds.

As Watson and Crick noted in their report, "It has not escaped our notice that the specific pairing we have postulated immediately suggests a possible copying mechanism for the genetic material." This pairing caused one side of the DNA strand to be the opposite image of the strand's other side. This reverse image then created the original image by attracting more base pairs to take the place of those that had been lost when the chromosome divided.

Adenine, for example, combines only with thymine. Therefore, when the DNA is split, the thymine has an open spot in which only adenine can fit. It will pair up with another molecule of adenine while the adenine on the other strand pairs up with another molecule of thymine. When all of the bases bond with the correct nitrogen base, the result will be two double helices of DNA that are identical to each other.

This process allows for endless possible variations. Genes can be made up of many thousands of base pairs. If any one of these pairs jumps out of sequence or is miscopied, or if an extra base pair appears or one falls out of the chain, the copy will vary and be slightly different from the original. This ability to duplicate with a slight chance of a tiny alteration provided the constantly changing source of variety that Charles Darwin believed existed.

Scientists eventually discovered that the function of DNA in a cell was slightly more complicated. It also involved a molecule similar to DNA, called RNA (ribonucleic acid), that transferred the sequence of base pairs from the cell nucleus during the formation of proteins.

George Beadle, another Nobel-winning geneticist, summed up the impact of Watson and Crick's discovery on the scientific community: "I regard the working out of the detailed structure of DNA as one of the great achievements of biology in the twentieth century."

The Nobel Prize committee members agreed, even though at the time they awarded the prize they could think of "no immediate practical application" for Watson and Crick's work. In reality, however, their discovery was even more significant than they themselves realized because stunning uses and scientific advances gained from their insights were right around the corner.

RNA (ribonucleic acid): a nucleic acid that in most cells transmits genetic information from DNA in the nucleus to the cytoplasm and functions in the formation of proteins

CHAPTER SIX

Har Gobind Khorana and Synthetic Genes

Once scientists understood how DNA (deoxyribonucleic acid) reproduced itself, many became intrigued with the possibility of manipulating DNA. It stood to reason that if a certain sequence of base pairs produced a particular trait, then scientists could alter that trait by rearranging that sequence. They might even be able to create their own genes by linking base pairs in different sequences.

But scientists first had to crack the DNA code. Unless they could understand what the sequences meant, they had no more chance of bringing about any meaningful change than a person who knows no English could write a book in that language. Har Gobind Khorana not only helped to learn the language of DNA, but he also used his knowledge to build an artificial gene.

Khorana says his birthdate is January 9, 1922, but admits that the date is only a guess. No birth

Har Gobind Khorana cracked one of the most difficult codes in the world: the DNA sequences.

records were kept in his home town of Raipur, a small village in India. Har was the youngest of Shri Ganput Rai and Shrimat Krishna Devi Khorana's five children. Ganput was a tax collector for the British government that ruled India at the time. Because the Khoranas were one of the few literate families in the town, they did not have much in common with their 100 or so village neighbors.

Most of Har's early schooling took place outside under a tree, where the village teacher conducted classes. He did well enough in his studies to advance to high school in the nearby city of Multan. He then attended Punjab University on a scholarship, where he obtained a degree in chemistry in 1943. After earning a master's degree in 1945, he traveled to England on a scholarship from the government of India to study for his doctorate in organic chemistry at Liverpool University.

Later in his life, Khorana said, "You stay intellectually alive longer if you change your environment every so often." That philosophy explained why he hopscotched around the globe during his career. After earning his Ph.D. degree in 1948, he spent a year studying in Switzerland. He then returned to England in 1950 to accept the Nuffield Fellowship at Cambridge University. There Khorana worked with Sir Alexander Todd, who later won a Nobel Prize in chemistry for his work with nucleic acids.

In 1952, Khorana married a Swiss woman, Esther Sibler, and they had three children. He continued his world travels, stopping in western

Canada to serve as director of organic chemistry at the University of British Columbia in Vancouver.

Although the university had only limited funds for his department, it allowed Khorana the freedom to pursue his own studies. He pulled together a top research team. In 1959, he won recognition for synthesizing coenzyme A, which is important in the chemical processes of converting proteins, sugars, and fats into usable forms in human cells. The enzyme had cost $17,000 per ounce to isolate from natural sources, but Khorana made its production practical by building it from inexpensive chemicals.

In search of another shot of intellectual stimulation, in 1960 Khorana moved to the University of Wisconsin at Madison. He brought along four of his assistants and began work on research involving DNA and the nature of genes.

Khorana was more than intellectually gifted. He was also tireless, almost fanatical, about his job. For 12 years, he worked long hours without taking a vacation. Even when Khorana was relaxing, his mind seldom wandered far from the laboratory. He enjoyed music and walks in the woods, partly because they would often bring out his best ideas, which he would jot down on colored index cards. In contrast to James Watson and Francis Crick, Khorana was shy and withdrawn and preferred to do his thinking alone rather than bounce ideas off others. Shunning publicity, he would announce his discoveries—many of which were breathtaking—at small, quiet conferences with no more drama than if he were reading the daily stock market report.

The Nobel Prize committee often withholds awards from scientists until later in their careers to guard against the possibility that their honorees will make a more noteworthy discovery than that for which the award has been given. In Khorana's case, they did not wait long enough.

During the 1960s, Khorana got involved in the detective work of deciphering the DNA code. Other scientists figured out that the base pairs of DNA were arranged in sequences of three, or triplets. Each of these triplets contained the genetic instructions required for 1 of 20 amino acids. Various combinations of amino acids linked together to form the proteins that are essential to life.

The key to the code was to figure out which triplet combinations of base pairs signaled which amino acids. This was complicated by the fact that the four bases involved could occur in 64 different combinations, even though only 20 amino acids existed.

Building upon techniques pioneered by scientists in the 1950s, Khorana invented a method for synthesizing DNA and RNA (ribonucleic acid). He was able to perform this so efficiently that he artificially assembled all 64 possible base pair triplet combinations. When he carefully traced which amino acid each of these combinations produced, he discovered that several different triplets could send the code for the same amino acid.

Thanks to Khorana and others who contributed to this discovery, scientists now understood the basic language of DNA. They knew exactly which sequence of base pairs could summon the specific amino acid that the cell needed in order to build a particular protein. For this work, Khorana shared the 1968 Nobel Prize in physiology and medicine with Americans Marshall W. Nirenberg and Robert W. Holley.

But Khorana's work with genes was only the beginning. Even while he accepted the prize, he was striving to do something that scientists from Charles Darwin to Oswald Avery could scarcely have imagined. He wanted to create a gene, the basic unit of heredity.

Har Gobind Khorana receiving his 1968 Nobel Prize from King Gustaf Adolf of Sweden

THE BREAKTHROUGH

Prior to Khorana, scientists had succeeded in duplicating genes simply by taking nature's tools and allowing them to do what they were designed to do. They unzipped strands of DNA and used one half to recreate the missing half. Khorana, however, wondered about the possibility of assembling all of the molecules in the gene from ordinary chemicals.

In order to build a gene, Khorana first needed to know the sequences of the base pairs in that gene. When scientists in the 1960s were able to describe the exact sequence of a small gene in a yeast cell, Khorana had his starting point. Unlike many human genes that can contain hundreds of thousands of base pairs, this yeast gene consisted of only 77 base pairs.

Even with the simplicity of this gene, Khorana faced a monumental task. In building a nucleic acid, the addition of each base involved a complicated set of chemical reactions. A scientist could hardly hope to reproduce each chemical reaction perfectly. Each loss of efficiency in a reaction would mean that the next reaction would start out that much short of the ideal. In order to have any hope of succeeding, Khorana would have to painstakingly purify his product at each stage of his gene-building.

Because the task was far too time-consuming for a scientist to undertake alone, Khorana employed 24 highly trained assistants to work with him on the project. Dividing the huge task of building a gene into more manageable portions, his assistants worked on creating short sections of single-strand DNA, each containing no more than 10 base pairs.

Khorana had no illusions about the challenge he faced. In a lecture in 1968, he predicted that he was probably several years away from a finished synthetic yeast gene.

During the project, Khorana found ways of working more efficiently. He developed a splicing technique that helped to fuse these DNA pieces together. By arranging the first five bases of the second chain as complements to the last five bases of the first chain, the bases would join together with their complements to form an overlapping section of double-stranded DNA. The third section would then begin with the first five complementary bases of the last five bases of the second chains and so on. Khorana also discovered enzymes that would assist the process of joining DNA pieces together.

Khorana beat his prediction. By spring 1970, his team had completed their assigned task. They had created and joined 15 segments of DNA that were identical in order to that of the yeast gene. On June 2 of that year, Khorana announced to the world his creation of a gene. The victory, however, was somewhat hollow. After all the work that he and his team had put into the project, they could not get the gene to function as a normal yeast cell gene.

The process of creating proteins from genetic information was far more complicated than simply following a recipe that called for a certain number of particular amino acids. Although the scientists could imitate the structure of a gene, they did not know

Har Gobind Khorana at work in his laboratory at the University of Wisconsin in Madison

enough about its regulating mechanism. Khorana suspected that something activated the gene so it would start creating its specified protein. Then something else must turn it off so it was not constantly creating protein.

Undaunted, Khorana went back to work. To recharge his energy, he moved his operation and most of his research team to the Massachusetts

Institute of Technology in Cambridge. For his second attempt at building a gene, he selected a slightly more complex target—a gene from a form of bacteria found in the human intestine: *Escherichia coli (E. coli)*. Scientists had detected a sequence of nearly 200 base pairs in this gene. Although it would be more difficult to create than the yeast gene, this gene offered an important advantage. It was essential in the transfer of the amino acid tyrosine to the cell's protein-manufacturing centers. Knowing that, Khorana would have a relatively easy time determining whether his created gene was doing what it was supposed to do.

As before, the work was long, slow, and often tedious. Khorana assigned each member of the team the task of putting together one and one-half segments of DNA. They built 40 single-strand segments of 10 to 12 base pairs each and spliced them as before. The breakthrough came in 1973 when Khorana discovered the "start" and "stop" signals in the gene. His team completed the assembly of the DNA and installed the control signals at either end of the gene. As before, they had created a gene identical in appearance to the real thing. But could they get this gene to function?

In 1976, Khorana's team was ready for the crucial test. They inserted their artificial gene into an *Escherichia coli* cell and waited to see what happened. Then they breathed a collective sigh of relief when the cell carried out the gene's instructions perfectly.

THE RESULT

Stories like Mary Shelley's Frankenstein *illustrate that humans have always been fascinated by the idea of creating artificial life. The work of Har Gobind Khorana and others may make it possible to actually achieve that goal.*

In his typical low-key fashion, Khorana announced the creation of a fully operational synthetic gene at a small conference of scientists in August 1976. The accomplishment brought questions from nervous observers. Was Khorana's creation of a gene the dawn of a terrifying new age? Had scientists taken their craft so far that by using common chemicals off the laboratory shelf they could now manufacture and control life like the fictional Dr. Frankenstein? Who

could control what scientists were doing with their highly specialized techniques?

Geneticists assured critics that their fears were unfounded. Khorana had merely duplicated a couple of simple genes in very simple organisms. Humans had thousands of genes, many more than yeast and bacteria, and each human gene was thousands of times more complex than the ones that Khorana had created. Furthermore, Khorana's experience had shown that living creatures were not made by simply throwing all of the base pairs together in the proper proportions. Instead, they were the result of complex interactions between all of the proteins dictated by all the triplet base pairs.

Now some critics asked the opposite question. Did Khorana accomplish anything worthwhile, or was his gene-creation just a showy technology stunt? They noted that Khorana had tied up the services of 24 highly skilled scientists for 9 years, so it had taken a total of more than 200 years of human labor just to create the simplest of genes. What practical purpose could exist in a process that took so much effort to produce so little benefit?

Khorana pointed out that the ability to produce genes opened up a whole new realm of possibilities that would be of immense value in further research. Using techniques developed by Khorana (which others quickly improved upon), scientists could create small, specific changes in DNA. By studying the effects of these changes, they could learn more about the nature of mutations and the inheritance of variability.

To those who worried about the consequences of scientific tampering with life, Khorana replied that such scientific work would be necessary to solve the environmental, economic, and health problems that face humanity. "I do have a basic faith that the survival of our civilization is not even going to be possible without the proper use of science."

This, in turn, could lead to wondrous advances in genetic engineering, or the ability to manipulate genes to create useful products. Geneticists could discover how to add or remove certain triplet base pairs to fine-tune genes that would perform a desired result. For example, they might be able to create a gene that would produce a medicine that had been difficult and expensive to isolate. By inserting that new gene into living organisms, they could then pass on the desired trait to future generations.

In creating a gene that functioned as a gene should, Khorana offered final proof that the scientific theory about the structure of the gene and the nature of heredity was correct. His work with the gene wrote the final chapter in the ancient quest to find the source of heredity and variation in living things. What scientists do with that knowledge is another story, one that they are only now beginning to write.

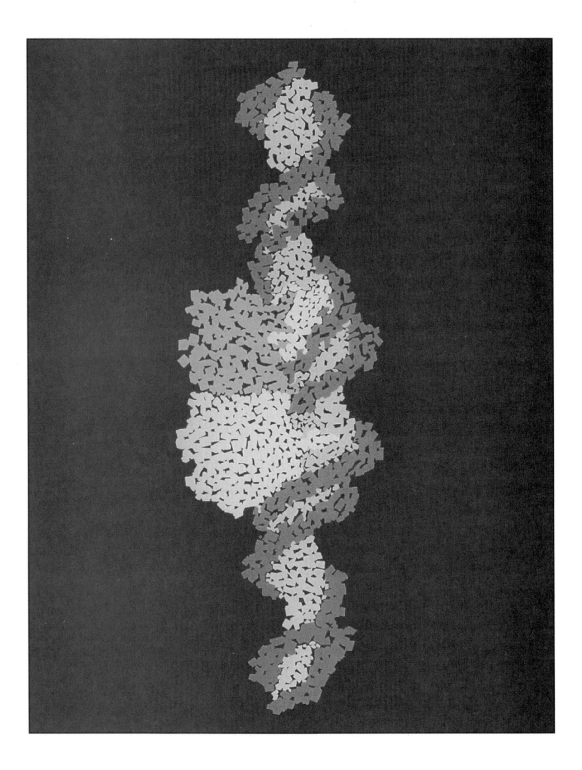

Shaping the World

On a cold November evening in 1983, Lynda Mann, age 15, walked to the village of Narborough, England, to visit a friend. She was found the next morning, raped and strangled. Police worked for three years to solve the case but got nowhere. In 1986, less than a mile from where the first murder had taken place, another 15-year-old, Dawn Ashworth, suffered the same fate as Mann. This time, police arrested a 17-year-old kitchen porter at a nearby hospital, who confessed to slaying Ashworth.

Curious as to whether the suspect was also involved in the Mann crime, police called on Alec Jeffreys to assist them. Jeffreys was a geneticist from the University of Leicester. He had been studying DNA in the chromosomes of various individuals, hoping to find "genetic markers"—something different in the genetic code of those who were diseased from those who were not afflicted. During this

The great discoveries in genetics during the last 100 years—such as the double helix structure of DNA— have done more than explain heredity. This new knowledge has become a tool for doctors, animal breeders, and even police detectives.

research, he saw that a section of DNA, called the intron, contained repetitive sequences of base pairs that appeared to be meaningless. The interesting thing about introns was that the number of repetitions seemed to be different in each individual.

Jeffreys and his assistants analyzed the DNA of a family to see if its members inherited the intron patterns. His test showed clearly that a person inherited half of the sequences from the mother and half from the father. Further tests found this identical DNA pattern in all parts of a person wherever nucleated cells are found—in hair, skin, semen, blood, saliva. When Jeffreys discovered that he could produce distinctive patterns from a single drop of blood, even if it were several years old, he realized that he had found a valuable tool for police work.

Jeffreys demonstrated his technique in a highly publicized immigration case. He proved that a young immigrant from Ghana was the son of a British subject and, therefore, was entitled to British citizenship. Intrigued by the possibilities of the identification technique that Jeffreys called "DNA fingerprinting," the police at Narborough asked the geneticist to help them in their case against the kitchen porter who had confessed to the 1986 slaying, but not to the earlier one.

Jeffreys obtained blood from the suspect and samples of the semen that had been left by the murderer at the scene of the two crimes. He then broke open the cells and isolated the DNA. Using enzymes, he cut out strands where the intron appeared. Then he placed the strands on a special gel plate and applied an electric current. This caused the DNA, which has a negative electrical charge, to move toward a charged pole at the other end of the plate. The smaller, lighter pieces moved faster, allowing him to sort the pieces by size.

To unzip the DNA, Jeffreys added another chemical. Finally, he added radioactive pieces of DNA that would lock on to his sample DNA wherever they found a complementary base pair. The radioactive pieces would show up on film, leaving an image of thin bands where the base pairs were located.

Jeffreys's test showed that the suspect in the murder of Dawn Ashworth was not the person who murdered Lynda Mann. Stunningly, it also showed

Professor Sir Alec Jeffreys with a DNA fingerprint produced by a technique he invented

that he had not murdered Ashworth either! The tests, however, indicated that the same individual had killed both girls.

Jeffreys's conclusion cost him the trust of many of the police officers on the case. But, lacking any important clues, the lead investigators stuck with the new technology and asked all young men in the village to submit to DNA testing. In January 1987, hundreds of men came forward to be tested and eliminated from suspicion.

After eight months of testing, police arrested a 27-year-old baker who had gotten a friend to give blood in his place. The friend had boasted in a local pub about deceiving the authorities. The DNA fingerprints of the baker, Colin Pitchfork, matched those of the murderer and he confessed to both crimes.

Thanks to Alec Jeffreys's DNA fingerprinting, a dangerous criminal was found and convicted, and an innocent man was set free. Since then, the technique has proved so useful in establishing guilt and innocence in cases where few other leads exist that many law enforcement officials consider DNA fingerprinting to be the most important advance in crimefighting in the past 100 years.

Speaking of the genetic research that led to his discovery of DNA fingerprinting, Jeffreys commented, "What we've really got is a new toy with endless possibilities and it's just amazing what could be done."

Amazing is almost too mild a word to describe what genetic researchers have been able to do in the last quarter of the twentieth century. Thanks to the discovery of "restriction enzymes" in the 1970s, the barriers to genetic engineering came crashing down. Led by Americans Herbert Boyer and Stanley Cohen, scientists discovered how they could use these enzymes to snip DNA into small pieces at precise locations. They then found other "recombinant" enzymes that attached DNA bits onto other DNA. With these tools, they could rearrange base pairs in strands of DNA, like editors splicing together a motion picture from several reels of film.

This accomplishment left scientists so in awe that they wondered whether they were trespassing into a realm that would be better left alone. Some questioned the ethics of experimenting with human genes, even for good purposes. Who would decide what was an "improvement" in the human genes? Legal questions concerning the creation of altered DNA were also raised. Could scientists claim patent rights for creating plants and animals with slightly altered genetic instructions? In 1973, scientists voluntarily stopped work on genetic engineering projects to assess the dangers of possibly producing new organisms that could devastate the world.

Within a few years, geneticists decided that the benefits of continuing research outweighed the risks. Scientific organizations removed almost all restrictions on genetic engineering research, and hundreds of companies organized to explore ways to create profitable and useful new forms of life. They also received a boost when the courts ruled in favor of genetic engineers.

A landmark case concerned Ananda Chakrabarty, a professor of biochemistry at the University of Illinois. Chakrabarty took bacteria that lived on oil and used their genes to develop a new strain of bacteria that could break down oil into simpler substances. Realizing how valuable this bacteria could be in cleaning up oil spills, he applied for a patent. The U.S. Patent Office turned him down, saying that life could not be patented. But on June 16, 1980, the U.S. Supreme Court ruled that scientists could patent genetically engineered living creatures.

The achievements in engineering came with astounding swiftness. In the 1970s, scientists spliced DNA into a bacterium that caused it to produce a

University of Illinois professor Ananda Chakrabarty developed a strain of bacteria capable of breaking down oil. The U.S. Supreme Court ruled in 1980 that he could patent this genetically engineered life form.

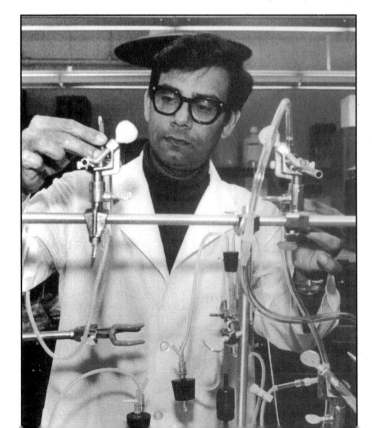

protein not previously found in that organism. In 1980, Martin Cline placed a gene from one mouse into another and showed that the gene continued to function after the transfer. A year later, researchers at Ohio University injected a rabbit gene into 312 mouse eggs. When they transplanted 211 of those eggs into the reproductive systems of mice, 46 of the eggs developed into live mice. When scientists mated two of the mice, five of the eight offspring also contained the rabbit gene. This proved that genes not only could be transferred into different species, but that the new animal could also pass on these genes to the next generation, ensuring that the new genes remained a permanent part of the gene pool.

That same year, Swiss scientists at the University of Geneva took genetic material from a mouse embryo and transplanted it into mouse eggs from which the genetic material had been removed. The result was a new mouse that was genetically identical to the mouse from which it had gained the genes. This process was called cloning. Chinese scientists followed up on this achievement by cloning a golden carp. Later in the decade, U.S. scientists cloned a sheep.

In 1982, geneticists inserted into a mouse a gene involved in the production of growth hormone. This produced a mouse twice the size of a normal mouse. By 1990, scientists were able to genetically engineer an entire mouse.

Meanwhile, large teams of scientists in France and the United States poured billions of dollars into an effort to catalog and map all the human genes. Their goal was eventually to determine the billions of bases that make up the estimated 20,000 to 100,000 genes on the 46 human chromosomes. This information would be especially valuable in determining the genetic causes of disease.

In the brief time since scientists have learned to manipulate genes, their work has already altered the world. Jeffreys's detective work is only one example of advances through genetic research. By comparing identical pieces of DNA from healthy and diseased persons, scientists have begun to uncover genetic markers that can alert doctors to the potential in people for developing certain diseases. They can also discover defective genes that might lead to future health-related problems. With advance warning of potential difficulties, doctors can treat at-risk people before the problem becomes more serious.

Genetic engineering has also allowed industries to produce valuable products far more cheaply. Insulin, for example, is a hormone crucial to the control of diabetes. Formerly, pharmaceutical companies obtained this substance from slaughtered animals. Now, however, genetic researchers have discovered how to insert a gene for insulin production into bacteria. Genes inserted in bacteria now pump out insulin cheaply and in great quantities as well as producing a variety of inexpensive chemical enzymes and other hormones.

Biologists use genetic fingerprints to find links between species and to see which species might have evolved from others. Environmentalists have used the same process in their fight to save endangered animals such as whales. Their DNA tests have proved that some whale meat on the market came from protected species. Historians have used DNA tests to identify the bones of Czar Nicholas II of Russia, putting to rest rumors that he had escaped execution in 1918. Anthropologists use DNA to study the bones of ancient humans to discover how modern ethnic groups are related to each other. Genetic research has also become an important part of agriculture. Gene manipulation has produced improved products such as tomatoes that stay edible for a longer period of time.

Nicholas II (1868-1918) was forced to abdicate his throne in 1917 during the Russian Revolution. He and his family were executed by the Communists in 1918.

Genetic engineering also holds promise for dramatic improvements in the quality of life. Targets include the creation of super plants that can ease world hunger, mass production of vaccines against disease, and the development of bacteria that can churn out industrial chemicals and other products without waste or pollution.

Manipulation of genes, however, particularly human genes, also holds the potential for destroying the dignity of human life. As Nobel Prize-winning biologist Salvador Luria has said, "When does a repaired or manufactured man stop being a man and become a robot, an object, an industrial product?"

Only time will tell what kind of a world Charles Darwin, Gregor Mendel, and the geneticists who followed them have created by unlocking the secrets of life.

NOBEL PRIZE WINNERS IN PHYSIOLOGY AND MEDICINE

Names in **bold** are people who are profiled in this book.
Names in ***bold italics*** are people who are mentioned in this book.

1901 Emil A. von Behring—serum therapy against diphtheria

1902 Ronald Ross—work on malaria

1903 Niels R. Finsen—treatment of lupus vulgaris, a rare form of tuberculosis

1904 Ivan P. Pavlov—physiology of digestion

1905 Robert Koch—work on tuberculosis

1906 Camillo Golgi and Santiago Ramón y Cajal—structure of the nervous system

1907 Charles L.A. Laveran—protozoa in disease

1908 Paul Ehrlich and Elie Metchnikoff—work on immunity

1909 Emil Theodor Kocher—work on thyroid gland

1910 Albrecht Kossel—chemistry of the cell

1911 Allvar Gullstrand—dioptrics (refraction of light) of the eye

1912 Alexis Carrel—grafting blood vessels and organs

1913 Charles Richet—work on hypersensitive reactions to foods or drugs

1914 Robert Bárány—physiology of vestibular system

1915-1918   NO AWARDS

1919 Jules Bordet—immunity

1920 August Krogh—regulation of the motor mechanism of capillaries

1921 NO AWARD

1922 Archibald V. Hill—heat production in muscles
Otto Meyerhof—oxygen use and lactic acid production in muscles

1923 Frederick G. Banting and John J. R. Macleod—discovery of insulin

1924 Willem Einthoven—invention of electrocardiogram

1925 NO AWARD

1926 Johannes Fibiger—discovery of Spiroptera carcinoma

1927 Julius Wagner-Jauregg—treatment of dementia paralytica

1928 Charles Nicolle—work on typhus exanthematicus

1929 Christiaan Eijkman—discovery of antineuritic vitamins
Frederick G. Hopkins—discovery of growth-promoting vitamins

1930   Karl Landsteiner—discovery of human blood groups

1931   Otto H. Warburg—respiratory work

1932   Charles Sherrington and Edgar D. Adrian—function of the neuron

1933   **Thomas H. Morgan**—discovery of hereditary function of chromosomes

1934   George H. Whipple and George R. Minot and William P. Murphy
       liver therapy against anemias

1935   Hans Spemann—embryonic development

1936   Henry H. Dale and Otto Loewi—transmission of nerve impulses

1937   Albert Szent-Györgyi von Nagyrapolt—biological combustion

1938   Corneille Heymans—respiration

1939   Gerhard Domagk—antibacterial effect of prontocilate

1940-1942   NO AWARDS

1943   Henrik Dam and Edward A. Doisy—analysis of Vitamin K

1944   Joseph Erlanger and Herbert Spencer Gasser—functions of nerve threads

1945   Alexander Fleming and Ernst Boris Chain and Howard W. Florey
       discovery of penicillin

1946   Herman J. Muller—effects of X rays on genes

1947   Carl F. Cori and Gerty T. Cori—animal starch metabolism
       Bernardo A. Houssay—study of pituitary

1948   Paul H. Mueller—insect-killing properties of DDT

1949   Walter Rudolph Hess—brain control of body
       Antonio Caetano de Abreu Freire Egas Moniz—brain operation

1950   Philip S. Hench and Edward C. Kendall and Tadeus Reichstein
       hormones of adrenal cortex

1951   Max Theiler—vaccine against yellow fever

1952   Selman A. Waksman—co-discovery of streptomycin

1953   Fritz A. Lipmann and Hans Adolph Krebs—studies of living cells

1954   John F. Enders and Thomas H. Weller and Frederick C. Robbins
       cultivation of polio virus

1955   Hugo Theorell—oxidation enzymes

1956   Dickinson W. Richards Jr. and André F. Cournand and Werner Forssmann
       new ways to treat heart disease

1957   Daniel Bovet—developed allergy and muscle relaxing drugs

1958   Joshua Lederberg—genetic mechanisms
        George W. Beadle and Edward L. Tatum—how genes transmit
        hereditary characteristics

1959   Severo Ochoa and Arthur Kornberg—compounds within chromosomes

1960   Macfarlane Burnet and Peter Brian Medawar
        discovery of acquired immunological tolerance

1961   Georg von Bekesy—mechanisms of stimulation within cochlea

1962   **James D. Watson** and **Maurice H. F. Wilkins** and **Francis H. C. Crick**
        determined DNA structure

1963   Alan Lloyd Hodgkin and Andrew Fielding Huxley and John Carew Eccles
        research on nerve cells

1964   Konrad E. Bloch and Feodor Lynen—cholesterol and fatty acid

1965   François Jacob and André Lwolff and Jacques Monod
        regulatory activities in body cells

1966   Charles Brenton Huggins—treatment of prostate cancer
        Francis Peyton Rous—discovered tumor-producing viruses

1967   Haldan K. Hartline and George Wald and Ragnar Granit
        work on the human eye

1968   **Robert W. Holley** and **Har Gobind Khorana** and **Marshall W. Nirenberg**
        studies of genetic code

1969   Max Delbrück and Alfred D. Hershey and **Salvador E. Luria**
        virus infection in living cells

1970   Julius Axelrod and Ulf S. von Euler and Bernard Katz
        transmission of nerve impulses

1971   Earl W. Sutherland Jr.—how hormones work

1972   Gerald M. Edelman and Rodney R. Porter—antibodies

1973   Karl von Frisch and Konrad Lorenz and Nikolaas Tinbergen
        behavior patterns

1974   George E. Palade and Christian de Duve and Albert Claude
        inner workings of living cells

1975   David Baltimore and Howard M. Temin and Renato Dulbecco
        interaction between tumor viruses and genetic material of the cell

1976   Baruch S. Blumberg and D. Carleton Gajdusek—infectious diseases

1977 Rosalyn S. Yalow and Roger C. L. Guillemin and Andrew V. Schally
role of hormones in the body

1978 Daniel Nathans and Hamilton O. Smith and Werner Arber
discovery of restriction enzymes

1979 Allan McLeod Cormack and Godfrey Newbold Hounsfield
developed computed axial tomography (CAT scan) X-ray

1980 Baruj Benacerraf and George D. Snell and Jean Dausset
how cell structures relate to organ transplants and diseases

1981 Roger W. Sperry and David H. Hubel and Torsten N. Wiesel
brain organization and functions

1982 Sune K. Bergström and Bengt I. Samuelsson and John R. Vane
research in prostaglandins, a hormonelike substance involved in illnesses

1983 ***Barbara McClintock***—discovery of mobile genes

1984 Cesar Milstein and Georges J. F. Kohler and Niels K. Jerne—immunology

1985 Michael S. Brown and Joseph L. Goldstein—cholesterol metabolism

1986 Rita Levi-Montalcini and ***Stanley Cohen***—substances that
influence cell growth

1987 Susumu Tonegawa—immunological defenses

1988 Gertrude B. Elion and George H. Hitchings and James Black
drug treatment

1989 J. Michael Bishop and Harold E. Varmus—theory of cancer development

1990 Joseph E. Murray and E. Donnall Thomas—pioneering work in transplants

1991 Erwin Neher and Bert Sakmann—developed a technique
called patch clamp

1992 Edmond H. Fischer and Edwin G. Kerb—work on enzymes

1993 Richard J. Roberts and Phillip A. Sharp—discovery of split genes

1994 Alfred Gilman and Martin Rodbell—discovered G proteins, a cellular
switch crucial to hundreds of processes in the human body

1995 Edward Lewis and Eric Wieschaus and Christiane Nüsslein-Volhard
how genes control embryonic development

adenine: an organic base that is an essential component of DNA and RNA

amino acid: an organic compound that is the basic building block from which proteins are synthesized in the cell

anesthesia: loss of sensation of pain induced by an anesthetic

anesthetic: an agent that causes a lack of sensation with or without loss of consciousness; from the Greek word *anaisthesia* which means "without feeling"

atom: the smallest unit of an element, consisting of electrons that surround a central, dense nucleus

axial: in botany, the arrangement of leaves around a central stem

bacteria: microorganisms that can cause disease

bacteriophage: a virus that infects and destroys certain bacteria

biochemistry: the study of chemical substances and vital processes in living organisms

biology: the study of life

botany: the study of plants

cell: smallest structural unit of an organism that is capable of independent function. It is made up of an outer membrane, the main mass (cytoplasm), and a nucleus.

chemistry: the study of the composition, structure, properties, and reactions of matter, especially on the atomic or molecular level

chromosome: a strand made up primarily of DNA in the nucleus of animal and plant cells that carries the genes and transmits hereditary information

cytoplasm: all of the substance of a cell outside the nucleus

cytosine: an organic base that is an essential component of DNA and RNA

DNA (deoxyribonucleic acid): a nucleic acid that carries the genetic information in the cell; it consists of two long chains of nucleotides twisted into a double helix and joined by hydrogen bonds; the sequence of bases in nucleotides determines individual hereditary characteristics

dominant trait: a genetically inherited trait that appears in an individual (as opposed to the recessive trait that is in an individual's genetic makeup, but does not appear)

Drosophila melanogaster: a type of fruit fly often used in genetic research

egg: the female reproductive cell

environment: the totality of circumstances or conditions surrounding an organism or group of organisms

enzyme: a protein produced in cells that speeds up the rate of biological and chemical reactions

Escherichia coli (E. coli): a form of bacteria found in the human intestine

evolution: the theory that groups of plants, animals, and other organisms change through time in structure and biological function due mainly to natural selection

extinct: no longer existing or living

fertilization: the union of the male reproductive cell and the female reproductive cell (sperm and egg in animals and plants)

fossil: the remains of an organism from a past geological age that became buried in the earth's crust

gene: a basic unit of inheritance that occupies a specific place on a chromosome and determines a particular characteristic in the organism

generation: the interval of time between the birth of the parents and the birth of their offspring

genetic engineering: scientific alteration of the genetic makeup in a living organism

genetic marker: a DNA sequence that is associated with a particular gene or trait and is used to indicate the presence of that gene or trait

geneticist: a scientist who studies heredity

genetics: the study of heredity

geology: the study of the origin, history, and structure of the earth

guanine: an organic base that is an essential component of DNA and RNA

habitat: the environment in which an organism normally lives

helix: a spiral form or structure

heredity: the genetic transmission of traits from parents to offspring; or, the totality of genetic traits transmitted from parents to offspring

hybrid: the offspring of two individuals of different species

hydrogen: an element that is one of the ingredients of water and many biological compounds; needed in the formation of carbohydrates, making it essential for living things

inherit: to receive a characteristic from parents through genetic transmission

intron: certain sections of DNA containing repetitive sequences of base pairs; repetitions vary in number and can be used to identify an organism

"jumping genes": mobile genes capable of moving to different locations on a chromosome and creating variations of individual characteristics

microbiology: the branch of biology that studies microorganisms and their effects on living organisms

microorganism: a living thing, such as bacteria, too small for humans to see without a microscope

mitosis: the process of cell division in which the cell divides into two genetically identical cells

molecular biology: the branch of biology that studies the formation, structure, and activity of molecules that are essential to life; often the study of genetic materials

molecule: the smallest particle into which an element or a compound can be divided without changing its chemical or physical properties

mutate: to change structure suddenly

mutation: a sudden structural change in a gene or chromosome of an organism

natural selection: the process in which the organisms best adapted to their environment tend to survive and pass on their genetic makeup to offspring while those more poorly adapted tend not to reproduce and die out

naturalist: someone who studies and describes natural objects, plants, and animals, especially their origins and interrelationships

nitrogen: a nonmetallic element that is in many organic compounds and all proteins; nitrogen is important in the formation of proteins and nucleic acids and is essential to all living organisms

Nobel Prize: a highly prestigious international award for achievement in the fields of physics, chemistry, physiology or medicine, literature, economics, and peace

nuclein: any of the substances present in the nucleus of a cell—mostly protein, phosphoric acids, and nucleic acids

nucleotide: a compound consisting of a base, a sugar, and a phosphate group; DNA and RNA are made up of nucleotides

nucleus: a spherical structure within the cell that controls the cell and its functions; the nucleus contains genetic information for the growth, maintenance, and reproduction of the organism

ovary: one of a pair of female sex organs in mammals that produces eggs

ovum: the female reproductive cell, or egg

phosphate: a salt or an ester (an organic compound formed from an organic acid and alcohol) of phosphoric acid

phosphorus: a nonmetallic element that is essential to the body for calcium, protein, and glucose metabolism in the human body

pneumococcus: a bacterium that is the most common cause of bacterial pneumonia (plural is pneumococci)

pneumonia: a disease that is caused by viruses, bacteria, or other microorganisms and is characterized by the swelling of the lungs

pollination (self- and cross-): the transfer of pollen from the stamen (the male reproductive organ) to the pistil (the female reproductive organ) of a flower

protein: any of a group of complex organic molecules that consists of chains of amino acids and includes substances, such as enzymes, hormones, and antibodies, that regulate body functions

radioactive: releasing radiation (energy in the form of rays or waves)

recessive trait: a genetically inherited trait that does not appear in the individual but is part of the individual's genetic makeup. The trait will appear only if paired with a similar recessive gene.

recombinant DNA (rDNA): a process in which scientists take genetic material from one organism and combine it with the DNA of another organism

recombinant enzyme: an enzyme that attaches pieces of DNA onto other bits of DNA; used in gene splicing

regenerate: to grow back

restriction enzyme: an enzyme that splits DNA at specific sites to produce separate fragments; used in gene-splicing

RNA (ribonucleic acid): a nucleic acid that in most cells transmits genetic information from DNA in the nucleus to the cytoplasm and functions in the formation of proteins

species: a fundamental category in the classification of living things consisting of related organisms that are capable of breeding with each other

sperm: the male reproductive cell

spontaneous generation: a discredited theory that microscopic living organisms such as bacteria simply appear. Until Louis Pasteur proved that microbes travel through the air, many people believed this theory.

synthesize: to combine to form a new, complex product

synthetic: made artificially, not of natural origin

terminal: in botany, the growth of leaves at the end of a stem, branch, or stalk

thymine: an organic base that is an essential part of DNA

traits (genetic): a genetically determined characteristic or condition

tuberculosis: a chronic infectious disease of humans and animals that is characterized by the formation of tubercles (swellings or lesions) in the lungs and other tissues of the body

tyrosine: an amino acid that is found in most proteins

zoology: the branch of biology that studies the structure, physiology, development, and classification of animals

BIBLIOGRAPHY

Allen, Garland E. *Thomas Hunt Morgan: The Man and His Science.*
New York: Princeton University, 1978.

Arnold, Caroline. *Genetics: From Mendel to Gene-Splicing.*
New York: Watts, 1986.

Asimov, Isaac. *Asimov's Biographical Encyclopedia of Science and Technology,*
2nd rev. ed. New York: Doubleday, 1982.

———. *Asimov's Chronology of Science and Discovery.* New York:
Harper & Row, 1989.

———. *How Did We Find Out About DNA?* New York: Walker, 1985.

Bowler, Peter J. *Charles Darwin: The Man and His Influence.*
Cambridge, Mass.: Blackwell, 1990.

Chedd, Graham. *The New Biology.* New York: Basic Books, 1972.

Crick, Francis. *What Mad Pursuit?* New York: Basic Books, 1988.

Dash, Joan. *The Triumph of Discovery.* Englewood Cliffs, N.J.:
Julian Messner, 1991.

Fox, Daniel, Marcia Meldrum, and Ira Rezak. *Nobel Prize Laureates in
Medicine or Physiology: A Biographical Dictionary.*
New York: Garland, 1990.

Franklin-Barbajosa, Cassandra. "The New Science of Identity."
National Geographic, May 1992, 112-123.

Gorman, Christine. "The Race to Map Our Genes." *Time,* February 8, 1993.

Hoagland, Mahlon. *Discovery: The Search for DNA's Secrets.* Boston: Houghton
Mifflin, 1981.

Judson, Horace Freeland. *The Eighth Day of Creation.* New York:
Simon & Schuster, 1979.

Keller, Evelyn Fox. *A Feeling for the Organism: The Life and Work of Barbara McClintock.* New York: W. H. Freeman & Co., 1983.

Lampton, Christopher. *DNA and the Creation of Life.* New York: ARCO, 1983.

———. *DNA Fingerprinting.* New York: Watts, 1991.

Lappe, Marc. *Broken Code.* San Francisco: Sierra Club, 1985.

Lee, Thomas. *The Promise and Perils of the New Biology.* New York: Plenum, 1993.

McCarty, Maclyn. *The Transforming Principle: Discovering that Genes Are Made of DNA.* New York: W. W. Norton, 1985.

McCuen, Gary E. *Manipulating Life.* Hudson, Wisc.: GEM, 1985.

Mayr, Ernst. *The Growth of Biological Thought.* Cambridge, Mass.: Belknap, 1982.

Miller, Jonathan. *Darwin for Beginners.* New York: Pantheon, 1982.

Modern Scientists and Engineers. New York: McGraw-Hill, 1980.

Moorehead, Alan. *Darwin and the Beagle.* New York: Harper & Row, 1969.

"More About Cloned Mice," *Science*, April 10, 1981.

Serafini, Anthony. *The Epic History of Biology.* New York: Plenum, 1993.

Shine, Ian and Sylvia Wrobel. *Thomas Hunt Morgan: Pioneers of Genetics.* Lexington, Ken.: University Press of Kentucky, 1976.

Sturtevant, Alfred H. *A History of Genetics.* New York: Harper & Row, 1965.

Wambaugh, Joseph. *The Blooding.* New York: Bantam, 1989.

Wasson, Tyler. *Nobel Prize Winners.* New York: H. W. Wilson, 1987.

Watson, James D. *The Double Helix.* New York: Antheneum, 1968.

ABOUT THE AUTHOR

Nathan Aaseng is an award-winning author of more than 100 books for young people on a wide range of subjects including science, history, biography, social issues, sports, and business. He wrote the **Great Decisions** series of books as well as *Great Justices of the Supreme Court, America's Third-Party Presidential Candidates,* and *Treacherous Traitors,* all published by The Oliver Press, Inc. Aaseng lives in Eau Claire, Wisconsin, with his wife and children.

PHOTO ACKNOWLEDGEMENTS

Archive Photos: pp. 36, 84, 87, 88, 97, 105, 110.
Bettmann Archive: pp. 6, 21, 25, 26, 52, 62, 66, 69, 75.
The British Council and Sir Alec Jeffreys: p. 117.
The Field Museum, GEO81011, Chicago: p. 12.
Famous 19th Century Faces, Art Direct Book Co.: pp. 22, 33, 38.
Library of Congress: pp. 9 (both), 10, 11, 13, 14, 16, 18, 27, 29, 32, 51, 54, 57 (both), 93, 123.
Minnesota Historical Society: p. 35.
National Library of Medicine: pp. 58, 73, 81 (both), 92, 100, 114.
Plants and Flowers, Dover Publications, Inc.: pp. 8, 43.
Rockefeller University Archives: pp. 70, 77, 83.
University of Wisconsin—Madison Archives: pp. 108, 112.
Wide World Photos: p. 120.